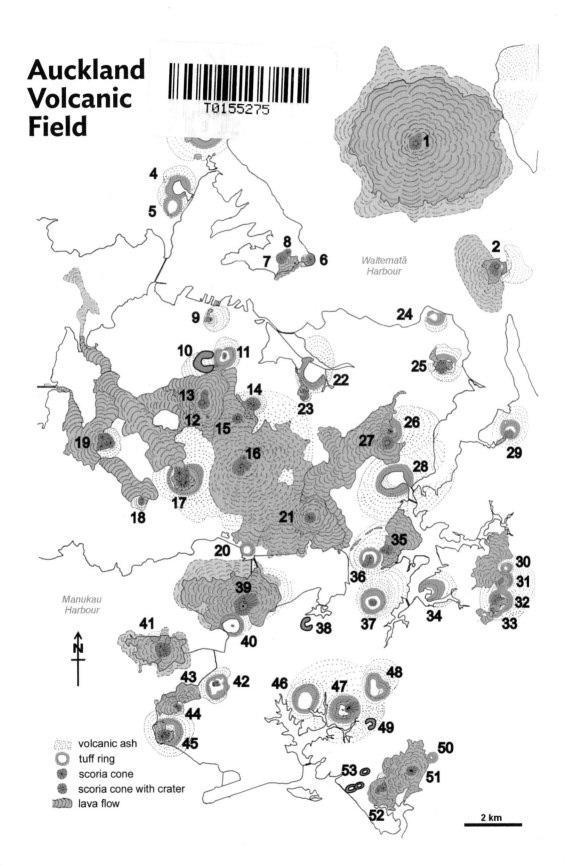

Auckland Volcanic Field

T0155275

Waitematā Harbour

Manukau Harbour

N

volcanic ash
tuff ring
scoria cone
scoria cone with crater
lava flow

2 km

Bruce W. Hayward is a retired geologist and marine ecologist based in Auckland. He is a former member of the Auckland Conservation Board and New Zealand Conservation Authority, and former president of the Geological Society of New Zealand. His wide interests in natural and human history have resulted in twenty previous books on topics as diverse as archaeology, the kauri timber and gum industries, the history of Auckland cinemas, New Zealand fossils, volcanoes, building stones and conservation; and, as joint author, the popular *A Field Guide to Auckland*.

Alastair Jamieson is an Auckland-based ecologist and photographer who has documented the region's changing natural environment with aerial photography for over 25 years. You can see more of his images at www.wildearthmedia.com

Volcanoes of Auckland
A Field Guide

Bruce W. Hayward

Aerial photography by
Alastair Jamieson

AUCKLAND
UNIVERSITY
PRESS

First published 2019
Auckland University Press
University of Auckland
Private Bag 92019
Auckland 1142
New Zealand
www.press.auckland.ac.nz

ISBN 978 1 86940 901 2

Published with the assistance of the Stout Trust

A catalogue record for this book is available from the National
Library of New Zealand

Design by Carolyn Lewis
Maps designed by Bruce W. Hayward

Front cover image: Eruptions some 60,000 years ago created
Maungakiekie/One Tree Hill, the second largest volcano in
the Auckland Volcanic Field and one of the most elaborate
archaeological site complexes in New Zealand. *Photo by
Alastair Jamieson*
Back cover image: The central scoria cone of Motukorea/Browns
Island, one of the least modified volcanoes remaining in the
Auckland Volcanic Field. *Photo by Alastair Jamieson*
Back endpaper image: Rangitoto Island's inhospitable lava flows
are still being colonised by a summer-flowering pōhutukawa
forest 600 years after the volcano ceased erupting. *Photo by
Alastair Jamieson*

Printed in China by Everbest Printing Investment Ltd

Contents

⊙ Maungarei/Mt Wellington scoria cone from the southwest in 2018. *Photo by Alastair Jamieson*

Introduction

Auckland is often referred to as the 'City of Sails' but it has an even stronger claim to the title 'City of Volcanoes'. Not only is much of the city built *over* the geologically young Auckland Volcanic Field, but much of the city is also built *out of* its volcanoes – the scoria cones and lava flows have provided much aggregate for constructing the city's buildings and its roads. The shape and form of the city's land area is strongly influenced by the landforms created by the volcanoes. Many of the city's hills and basins, as well as parts of the coastlines of both harbours, owe their origins to volcanic activity. Most of the remaining small volcanic cones are now much-loved parks that provide character to the city, playgrounds for its citizens and a sense of place for each suburb. Many of Auckland's older suburbs are named after their local volcano – Mt Eden, Mt Albert, Mt Roskill, Mt Wellington, Mt Richmond, Three Kings, One Tree Hill, Māngere, Ōtāhuhu, Remuera, Manurewa and Wiri.

Fifty-three volcanoes have erupted within 20 km of the city's centre, from Lake Pupuke in the north to Wiri Mountain in the south, and Mt Albert in the west to Pigeon Mountain in the east. For hundreds of years, these volcanoes have played a key part in the lives of Māori and Pākehā – as sites for Māori pā and 20th-century military fortifications, as kūmara gardens and parks, as sources of water and stone.

How many volcanoes are there? When did they erupt and how do we know? Will there be another eruption in Auckland and, if so, where and when? Will we have sufficient warning to evacuate in time? What is a lava cave, a volcanic bomb or a tuff ring? Why were Auckland's volcanoes such an attraction to early Māori? Why is it that Auckland's freshest water comes out of our volcanoes? This book sets out to answer these and many more questions.

The book has two parts – the first provides a general account of the geology and human history of Auckland's volcanoes, and the second, larger section provides a field guide for those who may wish to visit and explore each volcano or at least envisage what some of them once looked like before they were destroyed by our growing city.

Although this book is derived from *Volcanoes of Auckland: The Essential Guide,* published in 2011, its contents and format have been updated to reflect its purpose as a field guide. In the first part, a number of sections included in the previous book have been left out, whereas others, particularly those on the age of the volcanoes, have been extensively updated following recent research advances. In the second part, the sections on each of the volcanoes have been reworked to provide more detail about what you can see when visiting. Some illustrations have been retained but over 70 per cent of the photographs and maps are new. Since 2011, three additional volcanoes have been recognised (Boggust, Cemetery and Puhinui craters) and these are also included.

Auckland Volcanic Field

Most people think of a volcano as a single large cone that has built up by a series of eruptions over a lifetime of thousands to hundreds of thousands of years. All the iconic cone volcanoes of the world are of this kind and include most of New Zealand's best-known volcanoes, such as Mts Taranaki, Ruapehu, Ngāuruhoe and Whakaari/White Island. Over time, this type of volcano erupts a considerable amount of lava and ash, often from a single vent at the centre of the cone. They all have a magma chamber at a relatively shallow depth (5–10 km) beneath them that periodically clears its throat and erupts part of its contents at the surface.

Another type of volcanism occurs where there is no magma chamber in the crust but, periodically, relatively small quantities of molten mantle rock rise from depths of 60–90 km and erupt at the surface. Each of these eruptions surface in a different location and build a relatively small volcano that never erupts again. Volcanoes of this type erupt for a few days to several years and are separated by hundreds to thousands of years of inactivity. They tend to erupt within a discrete area that is called a volcanic field and the Auckland Volcanic Field is a classic example. In these volcanic fields the magma composition is consistently basalt. Volcanic fields with numerous small centres that each erupted once are called monogenetic. The exact number of volcanoes in the Auckland Volcanic Field depends on how they are counted, as some cones and craters are clustered close together and can be counted separately or in combination. This book recognises 53 individual volcanoes.

The Auckland Volcanic Field has erupted spasmodically over the last 200,000 years. While it is currently dormant, the last eruption,

◐ The best-known symbol or icon of Auckland City is Rangitoto Island – the youngest and largest volcano in the Auckland Volcanic Field.
◑ Motukorea/Browns Island is a combination of volcanic landforms (tuff ring, scoria cone, lava flows) produced by all three styles of eruption that characterise Auckland's volcanoes. *Photo by Alastair Jamieson*

Rangitoto, was only around 600 years ago and the field is considered to be still alive and likely to erupt again. All of Auckland's volcanoes, except Rangitoto, erupted on land. Sometimes the lava probably set the surrounding forest on fire but any evidence of this has disappeared. Each volcano would have killed the native forest in the immediate vicinity, and in several places the remains of these forests are buried and preserved by volcanic ash or lava flows. The two best and most easily viewed examples are at Takapuna Reef and at the end of Renton Rd, near Auckland Airport.

How the volcanoes work

The styles of eruption, types of rock produced and resulting landforms in Auckland's Volcanic Field

ERUPTION STYLE	SCIENTIFIC TERM	ROCK PRODUCED	LANDFORM
Wet explosive	Phreatomagmatic, Surtseyan & phreatic	Tuff (hardened volcanic ash), tuff breccia	Explosion crater (maar), tuff cone or tuff ring
Fire-fountaining & fiery explosive	Hawaiian & Strombolian	Scoria (lapilli, cinders), spatter, volcanic bombs	Scoria cone (cinder cone)
Lava outpouring	Strombolian & Hawaiian	Basaltic lava	Lava flow, lava field, or small lava shield

Wet explosive eruptions

When many of Auckland's volcanoes first erupted, the rising magma came into contact with near-surface groundwater in aquifers or swampy ground. Initially, the heat from the extremely hot magma (about 1000–1200 °C) caused the water to flash to steam, resulting in a violent explosion, rather like what happens when cold water is splashed into a pan of extremely hot oil. These first explosive eruptions may have involved only steam and are referred to as phreatic or steam eruptions. The steam blasted up through the overlying ground, throwing out large and small fragments of rock and soil and creating a small crater. Deposits from these kinds of eruption are often structureless heaps of angular blocks in a matrix of finely comminuted soil and rock (known as tuff breccia).

Commonly, but not always, the steam blast eruptions were followed by those that also involved the rising magma, the surface of which was instantly chilled, solidified and explosively fragmented when it encountered the cold water.

These eruptions are called phreatomagmatic as they involve both steam and magma. They result in the upwards and outwards ejection of a rapidly expanding cloud of steam, magmatic gas, fragmented lava and other pieces of rock from the vent walls.

The solid particles that erupted into the air and were deposited over the ground are referred to by the general term tephra. Tephra is divided on the basis of particle size into ash (fragments smaller than 2 mm), lapilli (2–64 mm across), and blocks and bombs (greater than 64 mm). Wet explosive eruption columns rose to heights of a kilometre or more and the less dense volcanic ash and lapilli within them were dispersed by the wind.

Fallout tephra accumulated on the ground on the downwind side of the volcano. Blocks of solid rock were ejected from the vent on ballistic trajectories and landed nearby. Around the denser base of the eruption column, base surges of superheated steam, gas, ash and lapilli were blasted out sideways at speeds up to 200 km/h.

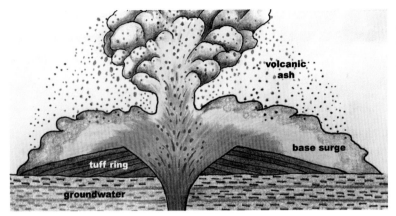

◐ Cartoon of a wet explosive eruption from an Auckland volcano. This style of eruption typically forms an explosion crater surrounded by a tuff ring composed of hardened layers of volcanic ash. *Illustration by Margaret Morley*

volcanic ash

base surge

tuff ring

groundwater

◑ Part of a sea cliff on Motukorea/Browns Island showing bedded tuff. Each layer represents an explosive pulse. Dark layers contain numerous grey basalt fragments derived from the rising magma and cream-coloured chunks of the underlying Waitematā Sandstone. Light-coloured laminated beds of fine ash were left behind by fast-moving base surges of searing gas, steam and ash. Height of photo 0.5 m. The black basalt fragment in the middle is a volcanic bomb in the impact crater where it landed in the soft ash.

These turbulent base surges devastated and partly buried areas within 3–5 km of the vent and were the most dangerous style of eruption produced by Auckland's volcanoes. As these surges passed they commonly left behind wavy beds of fine ash, sometimes with cross-bedded dune forms.

Wet explosive eruptions usually come in a series of pulsating episodes interspersed by short periods of inactivity. These eruptions typically produced a relatively shallow (50–100 m deep), wide (200–1000 m across), circular explosion crater surrounded by a low ring of bedded volcanic ash and lapilli. The ash and lapilli were erupted wet. As the layers dried out they hardened into a creamy-brown rock called tuff. The raised ring of tuff rock around the explosion crater is called a tuff ring, or, if it is a more substantial mound, it is sometimes called a tuff cone. A tuff ring, or cone, usually has its circular crest forming the rim of the crater with relatively steep slopes back into the crater and gentler slopes (*c.* 5–10 degrees) on the outside. The steeper inner slopes are often formed by a series of arcuate slump scarps as a result of sections of the tuff ring slipping back into the crater after being deposited.

If magma supply ceased before all the groundwater was used up, then the only land-form produced by the volcano was an explosion

crater surrounded by a tuff ring. There are 24 explosion craters with surrounding tuff rings in the Auckland Volcanic Field.

The rock fragments that were blasted out in these explosive eruptions include many pieces ripped from the walls of the volcano's throat. They range in size from small lapilli to large blocks more than a metre across. The composition of these ejected rocks provides information on the rock strata that underlie the volcano at depths of several hundred metres or more. In a way, these volcanoes were natural drilling rigs, bringing samples of rock to the surface for geologists to piece together and work out the underlying rock sequence.

After eruptions finished, the explosion craters gradually filled with rainwater, creating crater lakes. Many of these lakes have subsequently filled with sediment and are now tidal lagoons, swamps or reclaimed wetlands. Those that became tidal lagoons were formerly lakes that were breached by rising sea level about 8000 years ago as a result of the melting of the European and North American ice sheets after the end of the Last Ice Age.

Fire-fountaining and fiery explosive eruptions

If the water in the vent was all used up (during the wet explosive eruptions) before the magma supply waned, then eruptions switched to a dry style and scoria cones were built. These partly or completely filled the explosion crater and maybe even buried all trace of the tuff ring.

The magma that erupted to form Auckland's volcanoes was molten rock containing dissolved gas (mostly water vapour and carbon dioxide) under pressure, rather like the dissolved gas in a bottle of fizzy drink. As the rising magma neared the surface, pressure reduced and the releasing gas drove a fountaining of frothy liquid from the vent, called fire-fountaining. Shake up a bottle of Coca-Cola, remove the lid, partly close the opening with your thumb and watch the fountaining powered by the escaping gas – so it is with natural fire-fountaining.

As the fountaining frothy lava flew through the air, it cooled and solidified, forming the frothy rock known as scoria (sometimes called cinders). This rock is initially black but oxidation of iron (reaction with oxygen in the air) during

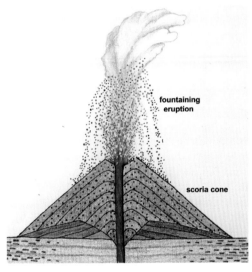

fountaining turned much of the hot scoria to its characteristic red colour. Scoria that remained black erupted on the margins of the fountain and cooled too quickly to react with the atmosphere.

The erupted scoria accumulated around the vent, building a steep-sided scoria cone with a deep crater. The slope of scoria cones is about 30 degrees, the angle at which scoria came to rest as it rolled downhill. A 30-metre-high scoria cone can be thrown up in less than a week and a 100-metre-high cone in several weeks. If a strong wind was blowing during fountaining, much of the scoria landed on the downwind side of the vent, building up a higher peak on one side of the scoria cone (e.g. One Tree Hill).

Scoria cones consist of layers of scoria of various sizes. Larger and denser fragments landed closer to the vent and the smallest lapilli and scoriaceous ash could be blown many kilometres away (e.g. Three Kings fountaining eruptions). Auckland's scoria cones were built mostly by steady fire-fountaining eruptions of rapidly rising, rather fluid magma containing multitudes of small gas bubbles. Sometimes this magma was still sufficiently liquid when it landed that it joined together to form a small lava flow that ran part-way down the cone slopes before cooling and solidifying. These small flows within a scoria cone are often called rootless lava flows because they differ from normal flows in that they have no obvious link to their source deep in the throat of the volcano.

Within some of the scoria cones there are layers of larger, ragged chunks of coarsely vesicular (holey) or more dense basalt that were expelled from the vent by discrete fiery explosive eruptions (called Strombolian style) of more pasty lava. These incandescent lumps often landed in a sticky molten form and tended to weld together into hard layers. The rate of magma ascent usually determines the style of eruption and the character of the ejected lava. The discrete fiery explosive outbursts occurred at fairly regular intervals seconds to minutes apart, and indicate more slowly rising, less fluid

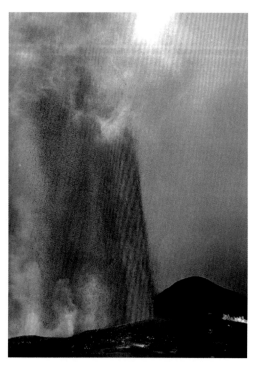

magma. Some of the small gas bubbles coalesced and grew extremely large as they rose through the magma column, each bursting at the surface to produce a separate explosive blast ejecting lumps of fiery lava tens to hundreds of metres into the air.

These discrete fiery explosive eruptions often occurred towards the end of the dry fire-fountaining phase as the rate of magma ascent slowed and it became cooler, thicker and less gaseous. As a result, a number of Auckland's scoria cones are capped by partly welded deposits of large, ragged lumps of scoriaceous basalt and aerodynamically shaped volcanic bombs (e.g. McLaughlins Mountain, Mt Wellington, Big King and Little Rangitoto).

Over half of Auckland's volcanoes produced one or more scoria cones erupted by a combination of fire-fountaining and fiery explosive eruptions, although many have been damaged or removed by quarrying in recent times. Several of the smallest cones (e.g. Pukeiti, Te Pou Hawaiki and Hampton Park) are composed almost entirely of the more welded deposits thrown out by fiery explosive eruptions.

◐ A fire-fountaining (Hawaiian) eruption in Hawai'i. Notice how the wind has blown the column so that the scoria lands on one side of the vent. *Courtesy of US Geological Survey*

◑ Well-sorted red scoria produced by fire-fountaining. The lumps of scoria are honeycombed with small holes (called vesicles) that were once bubbles in the erupting frothy lava.

Lava flows

During the dry fire-fountaining and fiery explosive phase of eruption of Auckland's scoria cones, the partly degassed molten magma often rose up inside the throat of the volcano. If it reached the height of the base of the scoria cone, this magma could push its way through the loose scoria and emerge as a flow of lava from near the base of the cone. Sometimes the loose scoria collapsed and the side of the scoria cone was rafted away by the outflowing lava. Such action created a horseshoe-shaped or breached crater (e.g. on Mt Hobson and One Tree Hill). Any scoria that landed on the lava flowing through the breach was also rafted away.

The outpouring of lava flows was usually accompanied by fire-fountaining or fiery explosions. Gas in the fluid magma was released in the volcano's throat and powered the fiery eruptions

⊙ Maungawhau/Mt Eden's steep-sided scoria cone was built by fire-fountaining and fiery explosive eruptions of bubbly lava (scoria) from two craters. The southern crater is the most obvious. *Photo by Alastair Jamieson, 2018*

from the vent directly above; whereas the lava that flowed out the side had lost most of its dissolved gas and, when it cooled and solidified, it became a relatively dense, dark grey basaltic rock. Some gas that was still trapped in the lava often rose towards the surface of the flow as it cooled. This sometimes resulted in a zone of more vesicular, or holey, basalt near the top of the flow.

As molten basaltic lava cools and solidifies, it contracts and cooling cracks form. These cracks often form fairly regular hexagonal-shaped columns (called columnar joints) that are vertical (or, more accurately, perpendicular to the cooling surfaces at the top and bottom of the flow). Sometimes, near-horizontal cooling joints also form near the top or bottom of a flow.

The size of the lava flow or field of coalescing lava flows depended on the supply of lava. Some of Auckland's cones produced just one small flow (e.g. Little Rangitoto and North Head), whereas others produced sizeable lava-flow fields that completely surrounded them (e.g. Māngere Mountain and Mt Eden). The speed and distance travelled by individual flows was controlled by the eruption rate and temperature of the erupted lava and hence its viscosity. The hotter, more liquid lava flowed downhill at running pace. Its surface chilled quickly to a thin, elastic black crust but the fluid lava beneath continued to flow and deform the surface skin into curved, ropey rolls, rather like the skin on a pot of cooling jam. Flows with such a smooth surface texture are known by their Hawaiian name as pahoehoe flows. As the flows moved downhill they cooled and became stickier, eventually stopping and solidifying into basalt rock.

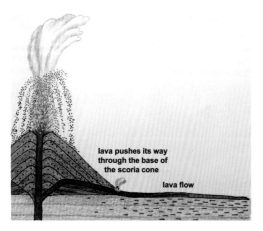

lava pushes its way
through the base of
the scoria cone

lava flow

⊙ Cartoon cross-section of the plumbing and eruption style that resulted in the outpouring of lava flows from the base of an Auckland scoria cone. *Illustration by Margaret Morley*

⊙ A pahoehoe lava flow in Hawai'i showing the characteristic cooled, black crust with its ropey rolls on the surface of the still molten lava. *Courtesy of US Geological Survey*

⊙ Typical rubbly surface of an a'a lava flow on Rangitoto Island.

Towards the end of a volcano's eruptive history, the magma in the plumbing began to cool and the erupted lava flows were more viscous. They flowed more slowly and often came to a standstill not far from the vents from which they oozed (e.g. One Tree Hill and Mt Eden). These cooler, more viscous lava flows formed thicker, more solid surface crusts that behaved brittly and not elastically. This thick crust was broken up into sharp blocks of rotating basalt by the continued movement of the fluid lava interior of the flow. These flows, which look like a moving pile of rocky rubble, are known by their Hawaiian name as a'a flows.

The two Auckland volcanoes that erupted the largest volume of lava are One Tree Hill and Rangitoto. Over several years, numerous flows poured out from around the base of their growing scoria cones. These flows cooled and solidified on top of each other to build up roughly circular cones with gentle slopes of about 10 degrees. These gently sloping lava-flow cones are small shield volcanoes – Rangitoto's is the best developed, and One Tree Hill's is less so and best viewed from the south. McLennan Hills also built up a small shield volcano. All three were capped by several coalescing, steeper scoria cones at the centre.

Where the volume of outpouring lava was less, the shape of the land over which it erupted influenced the resulting form of the lava-flow field. Flows that erupted onto the relatively flat Manukau lowlands generally spread out as a wide apron close to the central cone (e.g. Māngere Mountain and Ōtuataua). The Auckland Isthmus was more dissected with incised streams draining rolling sandstone hill country. In this area, outpouring lava flowed as viscous rivers down stream valleys and cooled to form solid ribbons of basalt filling the valley floors (e.g. the Three Kings and Mt Wellington flows). The streams were displaced and now flow along the side of the basaltic flows, eroding new courses into the softer sandstone banks (e.g. Oakley, Meola, Motions and Puhinui creeks).

◐ As lava cooled and solidified it contracted, forming large hexagonal columnar fractures like these within a thick lava flow from Mt Eden. This jointing assisted inmates from nearby Mt Eden Prison with the back-breaking job of splitting the basalt rock into blocks suitable for building and kerb stones. The jointed rock in the old quarry faces in Auckland Grammar School grounds now provides challenges for Auckland's rock climbers.

Volcanic bombs and projectile blocks

During fiery explosive eruptions of Auckland's volcanoes, magma ascent slowed and the rising magma became more pasty, or viscous. As a result, gas bubbles rose more slowly and many coalesced into massive balloons of gas sometimes many metres across. These giant gas balloons burst at the surface of the magma column with such force that large globs of magma were thrown high into the air on various ballistic trajectories out of the volcano's vent. During flight, this cooling plastic material often acquired rounded aerodynamic shapes before landing as semi-solid volcanic bombs. Most volcanic bombs are 6–30 cm in size, but occasionally they reach a metre or slightly more across. Volcanic bombs were often some of the last material thrown out from Auckland's scoria cones as the supply of rising magma waned, then cooled and became more viscous while still in the volcano's throat.

A second type of large projectile was some-times ejected from Auckland's volcanoes. These were angular blocks of solid rock, mostly thrown out by explosive eruptions. They are found in volcanic ash deposits (tuff) in the tuff rings or partly welded, scoriaceous layers in the scoria cones. They may be pieces of existing country rock (such as Waitematā Sandstone) that were ripped from the walls of the volcano's throat, or blocks of disrupted solid basalt that had cooled and solidified, perhaps on the surface of a lava lake during a lull in eruption activity.

◐ The bursting of large bubbles of gas as they reached the surface of molten magma inside a volcano's throat threw out globs of magma in the same way as a bursting bubble throws out globs of mud from Rotorua's geothermal mud pools. These molten globs developed various aerodynamic shapes as they flew and cooled through the air, before landing as near-solid masses called volcanic bombs.

⊙ Examples of different shapes of volcanic bomb that have been thrown out from Auckland's scoria cones: **A.** Spindle bombs – plastic globs of lava that spun during flight, producing elongate shapes tapering to tails at each end (also called fusiform bombs); **B.** Cowpat bombs – formed when hotter, less solid bombs landed splat on the ground, forming flattened disks like cowpats, Mt Wellington; **C.** Large fusiform bomb that was partly flattened as the still soft molten bomb hit the ground, Māngere Mountain.

◔ This 30 cm block of brown Waitematā Sandstone occurs within bedded tuff of the eroded Maungataketake tuff ring. The sandstone was ripped from the wall of the volcano's throat and fired out in an explosive eruption. Upon landing, its impact disrupted the layering as it buried itself 40 cm into the soft volcanic ash.

○ These small spheres of ash, called volcanic 'hailstones', form a layer within a sequence of ash deposits (tuff) from Maungataketake Volcano, Manukau Harbour.

Auckland's suburbs, but all are myths – the longest-known is Wiri Lava Cave at 290 m, and no cave can link between separate flows.

Accessible lava caves are uncommon. Only about 40 substantial lava caves are known in New Zealand and all occur within the basaltic lava flows from Auckland's volcanoes. Many of those that have been discovered have now been sealed off by buildings or streets, filled with rubbish or quarried away. The most easily visited lava caves are on Rangitoto Island, signposted from the main walking track up to the summit.

Volcanic 'hailstones'

Volcanic 'hailstones' are most commonly formed inside an intensely hot, steamy base surge eruption. In these, the fine volcanic ash grains may stick together with the moisture and form small balls 2–5 mm across. The balls fall out of the passing surge and are deposited in layers within the tuff ring. While volcanic hailstones is their common name, geologists call them accretionary lapilli or chalazoidites. They are most easily seen in tuff deposits from Three Kings and Maungataketake volcanoes.

Lava caves

Most lava caves form inside lava flows. As the molten lava flows away from the vent it cools rapidly on the outside of the flow to form a solid crust of grey basalt. The hot lava continues to move along inside this tube. The crust thickens from the inside as cooling lava adheres to it. As the supply ebbs, the lava core drains, leaving an empty tube, or lava cave. The roofs of these caves have collapsed in a number of places, forming 'skylights', and it is through these holes that people get access today. Because of roof collapses, most caves are not very long, with few longer than 50 m. Hearsay stories abound of lava caves stretching many kilometres beneath

○ Two cross-sections showing lava cave formation. Most lava caves form inside lava flows. The outside of the flow cools and solidifies, forming a crust of basalt around an internal tube filled with flowing hot lava. When the supply of lava stops, the molten liquid drains out leaving an empty tube or cave. *Illustrations by Geoff Cox*

Stewarts Lava Cave, in a flow from Three Kings Volcano, is another that is periodically visited by guided groups who obtain permission from the private owner of the cave's entrance.

The insides of the caves are not just featureless tubes. Many caves have small cylindrical blowholes through their roofs where trapped gas under pressure forced a hole through the soft lava. Some have short lava stalactites hanging from their roofs and lava dribbles down their upper walls, especially around the blow-holes. These were formed when extremely hot gas was trapped between the top of the flowing lava and the cave ceiling. The heat from the gas partially remelted the basalt lining giving it a glazed appearance. Occasionally, molten drops of lava fell from these stalactites and built up low stalagmites on the floor.

Cave walls often have shelf-like encrustations of solid lava, called surge benches. These are mostly formed at the high stands reached by successive surges of lava through the feeder tube and leave behind narrow benches of cooled lava that mark their maximum height on the cave wall. The cave floors usually consist of the dregs of the last surge of lava that flowed through the tube. Often these are rucked up into curved ropes of pahoehoe lava on their surface.

◔ Ambury Lava Cave inside a Māngere Mountain lava flow has a high roof and an artificially flattened floor. *Photo by Alastair Jamieson, 2010*

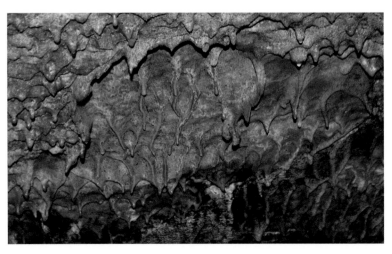

◔ Small lava stalactites hang from the ceiling of this Onehunga cave in a One Tree Hill lava flow.
◔ Locations and names of significant lava caves in Auckland.

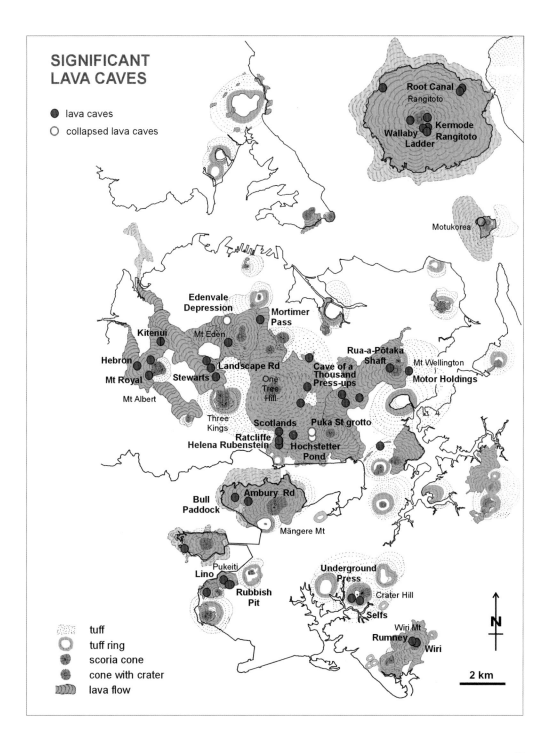

SIGNIFICANT LAVA CAVES

● lava caves
○ collapsed lava caves

Root Canal
Rangitoto
Kermode
Rangitoto
Wallaby
Ladder

Motukorea

Edenvale Depression
Mortimer Pass
Kitenui
Mt Eden
Rua-a-Pōtaka Shaft
Mt Wellington
Hebron
Landscape Rd
Cave of a Thousand Press-ups
Motor Holdings
Mt Royal
Stewarts
One Tree Hill
Mt Albert
Three Kings
Scotlands
Puka St grotto
Ratcliffe
Helena Rubenstein
Hochstetter Pond

Bull Paddock
Ambury Rd
Māngere Mt

Pukeiti
Lino
Rubbish Pit
Underground Press
Crater Hill
Selfs
Wiri Mt
Rumney
Wiri

tuff
tuff ring
scoria cone
cone with crater
lava flow

N

2 km

15

How old is each volcano?

There are a number of ways to determine the relative or absolute age of each of Auckland's volcanoes. Most methods are relatively new and have been developed and used only since the 1990s. Prior to that, it was thought that all of Auckland's volcanoes had erupted within only the last 50,000 years, but more recent studies have now extended the time span for eruptions back to the last 200,000 years.

Formerly, the majority of age determinations were by the radiocarbon dating method using pieces of fossil wood that were obtained by the late Professor Ernie Searle and others from beneath or within Auckland's lava flows and volcanic ash. These determinations, made in the 1950s and 1960s, mostly gave ages of about 30,000 years or less, and were accepted at face value to be correct. Recently it has become clear that 30,000 years ago was an approximate maximum limit that radiocarbon could date at the time and that many of these early dates are not reliable and more likely minimum ages.

In the last 25 years several new techniques, such as measurements of magnetic reversals, argon-argon (Ar-Ar) isotopic dating and geochemical fingerprinting of volcanic ash layers, have been used to obtain many new ages for Auckland's volcanoes. The table on pages 21–22 and the figure on page 25 document what we now know and are confident to accept about the age of Auckland's volcanoes. The youngest is Rangitoto at about 620–600 years old, and the oldest-known at about 200,000 years are Pupuke, Onepoto and Tuff Crater volcanoes on the North Shore.

Relative ages

Prior to the introduction of numerical dating methods there were two ways of determining the age of Auckland's volcanoes relative to one another. The first was to use the state of weathering and erosion of each volcano's rocks and landforms. By this method, Rangitoto clearly has the freshest rocks and is the youngest volcano in the field. Others have scoria cones or tuff rings that are obviously more weathered and eroded and are, therefore, amongst the older volcanoes. These included the Auckland Domain, Grafton, Glover Park, Boggust Park and Pukewairiki volcanoes. Pukewairiki tuff cone has a coastal terrace, 10 m ASL, eroded into its western side by the Tāmaki Estuary. Presumably this occurred when sea level was higher than at present during the Last Interglacial warm period about 130,000 years ago and, therefore, Pukewairiki is older than that. Similarly, Boggust Park has a wide-open breach in its tuff ring to the east, which was presumably eroded by tidal currents rushing in and out also during the Last Interglacial period of higher sea level.

The second way of telling the relative age of the volcanoes is where a geologist finds the erupted ash or lava flow from one volcano overlying or being diverted around the volcanic landforms of a neighbour. It is clear, for example, that Ōrākei Basin erupted before Little Rangitoto, as a lava flow from the latter flowed around the side of Ōrākei Basin tuff ring. In three places lava flows from one volcano are seen or used to be seen, prior to quarrying, to have flowed into the explosion crater of a pre-existing neighbouring volcano. Thus Mt Wellington erupted after Purchas Hill, Māngere Mountain after Māngere Lagoon and Ōtara Hill after Hampton Park Volcano. In these three instances, no soil or weathered layer has been seen between the lava flows and underlying ash and, therefore, the time gap between the neighbouring volcanoes may have been only short.

In drill holes near Eden Park, however, lava flows from Mt Eden overlie a lava flow that has been linked back to Mt St John and they are separated by up to 1 m of peat. This peat would have accumulated in a swamp and may represent tens of thousands of years between the two eruptions. On the Onehunga foreshore, the volcanic ash deposits erupted from Te Hopua explosion crater contain large blocks of basalt that have been linked to One Tree Hill eruptions, indicating that Te Hopua is younger than, and erupted through, the basaltic lava flows from One Tree Hill.

In some places it is possible to trace wind-blown volcanic ash deposits back to their source volcano, as the ash becomes thicker and coarser closer to the vent. By this method, the late Les Kermode showed that ash from Māngere Mountain and Mt Smart overlie One Tree Hill flows and, therefore, both were younger. At East Tāmaki, ash from Styaks Swamp explosion crater overlies Green Mount flows and, near the city centre, ash from Auckland Domain overlies scoria and basalt from Grafton Volcano. The most useful volcanic ash for helping determine the relative ages of Auckland's isthmus volcanoes came from the voluminous eruptions from Three Kings, which is dated at 28,500 years old. Three Kings ash drapes the scoria cones and lava flows from One Tree Hill, Mt St John and Mt Hobson, indicating that all three are older. Nearby, Mt Eden and its lava flows have no cover of Three Kings ash and thus Mt Eden is younger than 28,500 years.

Radiocarbon dating

Radiocarbon dating has been used for over 60 years to date samples of organic material, such as wood, charcoal and peat, that have been buried by ash or lava flows from Auckland's volcanoes. This dating method relies on the natural decay of the radioactive isotope of carbon, ^{14}C. The carbon-14 isotope is a tiny fraction (10^{-12}%) of all organic carbon when it is made via cosmic rays in the upper atmosphere and it gradually decays to the stable nitrogen isotope, ^{14}N. The half-life of ^{14}C is 5730 years, which means that it takes this long for half the original amount of ^{14}C

to naturally decay to the stable isotope, and a further 5730 years for half of the remainder, and so on. In this technique the quantities of the three isotopes of carbon in each sample are measured accurately using advanced mass spectrometry instruments and these are used together with the known half-life and original proportion of ^{14}C relative to stable isotopes to calculate a sample's age. Radiocarbon ages are usually cited as years Before Present (BP), where present is standardised as AD 1950.

By about 40,000–50,000 years after formation there is so little ^{14}C left, it is difficult to reliably measure the amount, and thus samples older than 50,000 years cannot be dated by this technique. The instrumental accuracy of radiocarbon dates varies depending upon the amount of radiocarbon present but today is usually about plus or minus 50 years for samples between 400 and 1000 years old, to plus or minus 200–500 years for samples around 20,000 years old. The age of the tree or peat prior to its being killed or buried by the lava flow or volcanic ash needs also to be considered when determining the time of eruption. Organic material suitable for radiocarbon dating has not yet been found for many of Auckland's volcanoes.

Argon-argon dating

Another isotopic method has been used recently by GNS Science volcanologist Graham Leonard to date Auckland's volcanoes. Unlike radiocarbon dating, which requires fossilised organic material for dating, argon-argon (^{40}Ar-^{39}Ar) dating measures small quantities of argon isotopes in samples of the basaltic lava itself. A complicated instrumental technique is used to measure the small amounts of ^{40}Ar and ^{39}Ar (produced by irradiation in a nuclear reactor as a proxy for the amount of potassium, K) in the finely crystalline groundmass of the basalt rock. Because of the moderately long half-life, this technique is better suited to volcanoes that are older than 40,000–50,000 years. There are currently instrumental limits on the accuracy of this method of around plus or minus 2000–8000 years for lavas 30,000–100,000 years old, and plus or minus 10,000–30,000 years for lavas around 200,000 years old.

Unfortunately, with rocks this young, this method is based on measurements of very small quantities of isotopes and also relies on having fresh dense basalt samples. Thus it is not always possible to be sure of the reliability of the determined Ar-Ar date and it needs to be checked against relative dating methods and the new fingerprinting ages recently obtained. In some instances the Ar-Ar dates are inconsistent with the fingerprinting age, which is usually considered to be more reliable.

Magnetic and paleomagnetic methods

Former University of Auckland geophysicists John Cassidy and Corinne Locke undertook an aeromagnetic survey flown back and forth at 450 m elevation above the Auckland Volcanic Field. Magnetic anomalies in the rocks beneath Auckland were detected. These reflect the strength of the magnetisation in the iron-rich rocks of Auckland's volcanoes. The survey showed that most volcanoes had positive magnetic anomalies, as would be produced in rocks erupting today or for most of the last 780,000 years. Surprisingly, however, eight volcanoes had unusual or negative magnetic signatures, indicating that they erupted during one or more of the few short intervals of dramatically decreased intensity in the Earth's magnetic field in the last 100,000 years.

The paleomagnetic method has also been applied to Auckland basalts. This is based on measuring the magnetic orientation and intensity of iron-bearing minerals that were captured in the basaltic lava when it solidified. As the lava cooled and crystallised, the iron-bearing minerals aligned themselves in the direction of the Earth's magnetic field at that time. The direction and strength of the Earth's magnetic field has varied through time, with some changes – called 'excursions' – being very short-lived.

These measurements indicate that the eight centres with unusual aeromagnetic signatures erupted during three magnetic excursions, two of which are well known from around the world and have been dated at 39,000–41,000 years ago (McLennan Hills) and ~30,000 years ago (Wiri, Crater Hill, Mt Richmond, Taylors Hill and Puketūtū). The time of the third magnetic excursion, during which neighbouring East Tāmaki volcanoes (Hampton Park and Ōtara Hill) erupted, has not been accurately determined.

Coring sediment in crater lakes

Since the late 1990s, a drilling programme led by geologists from the University of Auckland and GNS Science has retrieved cores of sediment that had accumulated on the floors of 10 lakes, marine lagoons or swamps that formed in Auckland's explosion craters. The lakes and lagoons are Lake Pupuke and Onepoto on the North Shore, Ōrākei and Panmure basins, and Te Hopua and Pūkaki lagoons. Drill holes in the swamp sediment of Auckland Domain, Glover Park, Waitomokia and Kohuora explosion craters provided much shorter and poorer records. The age of the sediment filling the bottom of the crater provides a minimum age for the time of eruption, as the lake may not have formed immediately after the crater was formed. The cored sediment that accumulated when the craters were freshwater lakes is usually laminated diatom-rich mud. Diatoms are freshwater algae that live in the lake water and have microscopic skeletons made of silica that fall to the lake floor when they die. This lake sediment accumulated slowly at rates of around 20 cm to 1 m per 1000 years. Many of these lake-filled explosion craters were breached by rising sea level as the world warmed up after the Last Ice Age. Breaching occurred between 9000 and 7500 years ago, after which the craters became tidal lagoons. In these lagoons, unlayered blue-grey shelly mud rapidly accumulated at rates of around 5–20 m per 1000 years.

Within the lake floor sediment there are scattered layers of wind-blown volcanic ash

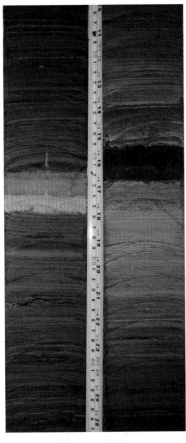

⊙ Sediment cores are extruded from the hydraulic piston coring barrel during drilling of the sediment sequence that accumulated in the Auckland Domain explosion crater.
⊙ The sediment infill in Lake Pupuke and Ōrākei and Panmure basins was cored using this drilling rig mounted on a barge to float on the lake or tidal lagoon waters. *Photo by Elaine Smid, DEVORA*

⊙ Two 40-centimetre-long lengths of drill core from Ōrākei Basin split open to display the laminated lake sediment and a 4-centimetre-thick cream rhyolitic ash bed (left core) erupted from the Taupō region and a 4-centimetre-thick black basaltic ash bed (right core) from an Auckland volcano. *Photos by Phil Shane*

of different thicknesses and compositions. Chemical and crystal analyses have been undertaken by specialist geologists called tephrochronologists, such as University of Auckland's Phil Shane and graduate students Joy Hoverd and Catherine Molloy. These studies have identified those ashes that are basalt erupted from Auckland's volcanoes, those that are andesite erupted from Mts Taranaki, Tongariro and Ruapehu, and those that are rhyolite erupted from the large Taupō, Ōkataina and Mayor Island (Tūhua) volcanic centres in the central North Island and Bay of Plenty. The rhyolitic ash deposits have been studied in great detail around the central North Island and many can be individually identified based on their chemical composition. A number have been reliably dated elsewhere, and these ages have been used to date the lake sediment in Auckland's craters. The ash layers provide the age framework for the sediment infill and to establish the sediment accumulation rates, which can be used to estimate the age of the oldest sediment in each crater.

Geochemical fingerprinting of volcanic ash layers

Unlike the andesitic and rhyolitic ashes from volcanoes well south of the city, the basaltic ash from Auckland's eruptions was blasted only a few kilometres into the air and was therefore not blown far from their vents. So, while most rhyolitic and andesitic ashes were deposited in every Auckland crater lake in existence at the time, Auckland's basaltic ashes occur only in those nearby or downwind from the source. Thus, a complete sequence of basaltic ash layers erupted from each of Auckland's volcanoes in turn is not present in any of the crater lakes. Instead, a composite sequence has had to be compiled by combining the records from the deeper cores taken from Pupuke, Onepoto, Ōrākei Basin, Te Hopua and Pūkaki Lagoon.

In the most recent breakthrough research, Victoria University geochemist Jenni Hopkins has compared the detailed chemical composition of these basaltic ashes to assist in recognising the same ash layers in different lake sediment cores. She has also compared the composition of the ash layers with that of tuff, scoria and basaltic lava of individual volcanoes. This is not a simple exercise, as the composition of the erupted lava changed through the course of each volcano's eruption, particularly in the longer-lived volcanoes.

Through this research Hopkins has managed to geochemically fingerprint the composition of each of the 28 separate basaltic ash layers she has identified in the crater-lake sediment sequences and identify one or several potential source volcanoes for most of them. For some ash layers the source volcano is unequivocally identifiable and thus she is confident that the core-based age of eruption of that volcano is correct, within the accuracy limits of sedimentation rates. These sedimentation rates are calculated above, below and between the well-dated rhyolitic ash layers in the cores, especially the Rotoehu Ash (~46,000 years old) and Ōruanui Ash (~25,000 years old). For other Auckland basaltic ash layers there are several possible source volcanoes based on their geochemical similarity. Sometimes the likely source volcano can be inferred because the ash layer is much thicker in one crater core than all others and this indicates that the source volcano was nearby.

Flare-up period

One of the most noticeable aspects of these new and reassessed dates is that there was a flare-up with the eruption of 10 volcanoes in a 3000-year period between ~31,000 and ~28,000 years ago. As noted earlier, this period includes five volcanoes with unusual negative magnetic signatures which indicate that they erupted during the Mono Lake magnetic excursion spanning a 600–1000-year period around 30,000 years ago. In addition to these five volcanoes, there are another five with normal positive magnetic signatures that have been dated as erupting just before or after the excursion. Eight separate basaltic ash layers occur in this time span in the crater-lake sediment sequences and seven of these have been correlated to their source volcanoes in the following order of eruption (from oldest to youngest): Crater Hill (~30,400 years), Mt Richmond, Taylors Hill (~30,200 years), Puketūtū (~29,800 years), Three Kings (~28,500 years) and Te Pou Hawaiki and Mt Eden (~28,000 years). No ash sourced from Wiri Mountain (Mono Lake anomaly age) or Ash Hill (radiocarbon dated at 32,000 years) has been found in any of the sediment sequences. This is not unexpected since they are both located in the far south and downwind of the craters that have been cored.

When the five volcanoes with a Mono Lake excursion signature were first recognised it was hypothesised that they all could potentially have erupted together from one batch of rising magma. The sedimentary core fingerprinting research shows that at least four of them erupted separately, each several hundred years apart, and all had different magma chemistries. Thus it seems likely that none erupted together; rather there were many eruptions in a short period of time.

The recently reassessed and new ages (years before today), allowing for accuracy limits, of Auckland's volcanoes

1.	Rangitoto	**620–600 C,L**
2.	Motukorea/Browns Island	**24,400±500 L**
3.	Pupuke Moana/Pupuke Volcano	~190,000 A
4.	Te Kopua-o-Matakamokamo/Tank Farm/Tuff Crater	~180,000 L
5.	Te Kopua-o-Matakerepo/Onepoto Basin	~185,000 L
6.	Maungauika/North Head	~90,000 A
7.	Takarunga/Mt Victoria	35,000±2000 L
8.	Takararo/Mt Cambria	~40,000 A
9.	Albert Park Volcano	~145,000 A
10.	Grafton Volcano	~100,000 L
11.	Pukekawa/Auckland Domain	~100,000 L
12.	Te Pou Hawaiki	~28,000? C
13.	Maungawhau/Mt Eden	**~28,000 L**
14.	Ōhinerangi/Mt Hobson/Ōhinerau	**34,000±1000 L**
15.	Te Kōpuke/Tītīkōpuke/Mt St John	~75,000 A
16.	Maungakiekie/One Tree Hill	**~60,000 L**
17.	Te Tātua-a-Riukiuta/Three Kings	**28,500±500 C,L**
18.	Puketāpapa/Pukewīwī/Mt Roskill	~105,000 A
19.	Te Ahi-kā-a-Rakataura/Ōwairaka/Mt Albert	~120,000 A
20.	Te Hopua-a-Rangi/Gloucester Park	>20,000 S
21.	Rarotonga/Mt Smart	**20,000±200 L**
22.	Ōrākei Basin	~120,000 L
23.	Maungarahiri/Little Rangitoto	**~24,500 A,L**
24.	Whakamuhu/Glover Park/St Heliers	160,000±20,000 L
25.	Taurere/Taylors Hill	**~30,000 L,M**
26.	Te Tauoma/Purchas Hill	**~10,000±200 C,L**
27.	Maungarei/Mt Wellington	**10,000±200 C,L**
28.	Te Kopua Kai-a-Hiku/Panmure Basin	**25,000±1000 L**
29.	Ōhuiarangi/Pigeon Mountain	24,500±500 L
30.	Styaks Swamp Crater	~20,000 A
31.	Matanginui/Green Mount	~20,000 A
32.	Te Puke-o-Taramainuku/Ōtara Hill	? M
33.	Hampton Park Volcano	? M
34.	Pukewairiki/Highbrook Park	>130,000 B

35.	Te Apunga-o-Tainui/McLennan Hills	**41,000±1000 A,C,M**
36.	Ōtāhuhu/Mt Richmond	**~30,000 L,M**
37.	Mt Robertson/Sturges Park	~24,000 L
38.	Boggust Park Crater	>130,000 B
39.	Te Pane-o-Mataaho/Māngere Mountain	**~50,000 L**
40.	Māngere Lagoon	**~50,000 L**
41.	Te Motu-a-Hiaroa/Puketūtū	**~30,000 A,C,L,M**
42.	Moerangi/Waitomokia/Mt Gabriel	**20,300±2000 L**
43.	Te Puketapapakanga-a-Hape/Pukeiti	~15,000 A
44.	Ōtuataua	~15,000 L
45.	Maungataketake/Elletts Mountain	~90,000 A
46.	Te Pūkaki Tapu-o-Poutukeka/Pūkaki Lagoon	>54,000 S
47.	Crater Hill	**~30,500 A,C,M,S**
48.	Kohuora Crater	**34,000±1000 C,L,S**
49.	Cemetery Crater	No data
50.	Ash Hill Crater	~30,500 C
51.	Te Manurewa-o-Tamapahore/Matukutūruru/Wiri Mountain	**~30,000 A,C,M**
52.	Matukutūreia/McLaughlins Mountain	~50,000 A
53.	Puhinui Craters	~50,000?

The reassessed ages in this table have been determined by: A Ar-Ar radiometric dating; B breached by Last Interglacial high sea level, 120,000–130,000 years ago; C radiocarbon dating (14C) of wood, peat or shell; L geochemical correlation of basaltic ashes in lake sediment cores with source volcanoes; M paleomagnetic method; S minimum age based on the oldest lake sediment recovered from their explosion craters. Dates have been sourced, with minor modifications, from Hopkins et al. (2017), Leonard et al. (2017) and Lindsay et al. (2011). (± = plus and minus limits of accuracy; > = older than; ~ = approximately.) Bolded ages are considered to be the most reliable.

Paired eruptions

Examination of exposures of the contacts between the volcanic material erupted from one volcano and another close by suggests that some volcanoes erupted soon after their neighbour with no time in between for any forest to become established or soil to develop. Indeed, with some of these volcanoes it seems likely that the eruption stopped at one vent and soon afterwards broke out within a few hundred metres at a new vent. This is probably because the first vent became blocked by a collapse of material back into the vent or by congealed magma.

Maybe a pause of several weeks to several years between separate batches or pulses of magma arriving at the surface was sufficient time for the magma still near the surface in the first vent to cool and solidify, creating a plug. The later magma that rose up the same pathway from the mantle towards the surface encountered the blockage and deviated to erupt nearby. In many cases these paired volcanoes are aligned in a near north–south axis. This suggests that near the surface the magma had pushed its way up a north-trending fault of broken-up rock and, when blocked, the magma simply injected itself

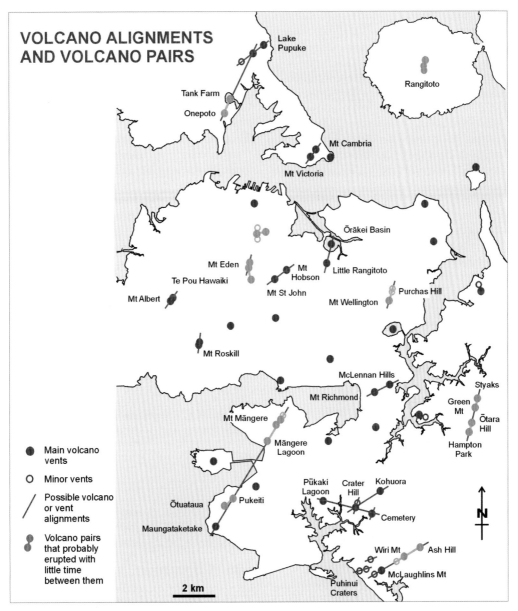

VOLCANO ALIGNMENTS AND VOLCANO PAIRS

Lake Pupuke
Rangitoto
Tank Farm
Onepoto
Mt Cambria
Mt Victoria
Ōrākei Basin
Mt Eden
Mt Hobson
Little Rangitoto
Te Pou Hawaiki
Mt St John
Purchas Hill
Mt Albert
Mt Wellington
Mt Roskill
McLennan Hills
Styaks
Mt Richmond
Green Mt
Mt Māngere
Ōtara Hill
Māngere Lagoon
Hampton Park
Pūkaki Lagoon
Crater Hill
Kohuora
Ōtuataua
Pukeiti
Cemetery
Maungataketake
Wiri Mt
Ash Hill
McLaughlins Mt
Puhinui Craters

Legend:
- Main volcano vents
- ○ Minor vents
- / Possible volcano or vent alignments
- Volcano pairs that probably erupted with little time between them

2 km

N

Map showing possible alignments of Auckland's volcanoes and vents that probably reflect deep underground faults which the batches of magma squeezed up along as they rose towards the surface to erupt. Also shown are probable pairs of volcanoes that appear to have erupted one straight after the other or with a small time gap of years to decades between. The magma for both volcanoes in the pair probably came up the same pathway from the mantle. Magma feeding the second volcano probably deviated sideways along a fault plane when it encountered a blockage in the plumbing (such as a solidified plug) near the surface.

sideways along the fault until it found a new way to the surface.

The possible paired eruptions include (from oldest to youngest): Onepoto–Tank Farm; Grafton–Auckland Domain, Māngere Lagoon–Māngere Mountain, Hampton Park–Ōtara Hill, Ash Hill–Wiri, Te Pou Hawaiki–Maungawhau, Pukeiti–Ōtuataua, Green Mount–Styaks Swamp, Purchas Hill–Maungarei and early Rangitoto–late Rangitoto. If each pair erupted close together in time as well as space, then maybe they should be considered as just one volcano. We have plenty of evidence of there being multiple vents or of eruptions moving from one vent to another within what we recognise as a single volcano. The only difference between multiple vents within one volcano and the paired volcanoes may be the distance between the various vents. Examples of volcanoes with more than one vent in an explosion crater include Pupuke, Waitomokia, Kohuora and Crater Hill. Examples of volcanoes with more than one fountaining vent and scoria cone include Maungawhau, Mt Roskill, Maungakiekie, Maungarei, Mt Richmond, McLennan Hills, Puketūtū, Three Kings and Māngere Mountain.

Age of Rangitoto eruptions

Reports in the media of evidence that Rangitoto erupted on and off for as long as 6000 years ago are based on contentious interpretations of several cores from Lake Pupuke and Rangitoto. The lava in the Rangitoto core that was dated at 6000 years old has been shown recently to be identical to the flows that erupted about 620 years ago and is now inferred to have been pushed into the soft sea floor sediment at that time. There is no debate that all the lava flows and scoria cones that form present-day Rangitoto were erupted in a short period of time (less than 20 years) and were witnessed by Māori on neighbouring Motutapu Island about 620–600 years ago. Ages of 550 and 500 years are often quoted in the media but these are the radiocarbon ages that are given as Before Present

(i.e. before 1950). The most accurate ages for the Rangitoto eruption come from cores taken from Motutapu Island peat swamps and Pupuke lake sediment. These give the age for the two phases of Rangitoto Island eruptions as about 620–600 years ago (AD ~1400–1420). It seems highly unlikely that Rangitoto ever erupted prior to this.

Eruptions and sea level

The Auckland Volcanic Field has been erupting intermittently over the last 200,000 years and during this time there have been two complete ice age climate cycles of warm and cold temperatures. Studies on fossil pollen indicate that during these cycles the Auckland area remained forested with no ice caps or glaciers, although the forest

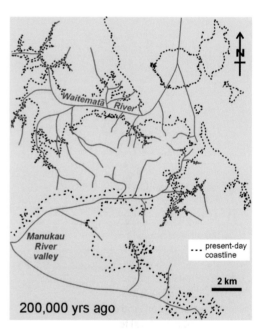

❶ Map of Auckland 200,000 years ago, before the volcanoes erupted. All but the youngest of Auckland's volcanoes erupted over a forested landscape of rolling hills and small valleys with streams feeding the eastern Waitematā or western Manukau river systems. For over 90 per cent of the period since 200,000 years ago, sea level was at least 30 m lower than present with no Waitematā or Manukau harbours.

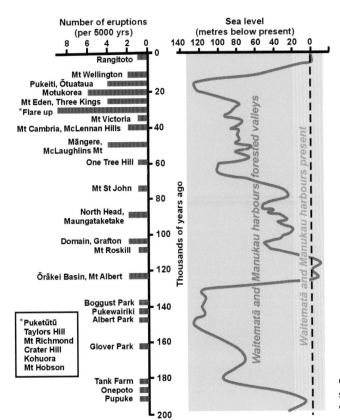

Number of eruptions (per 5000 yrs)

8 6 4 2 0

Sea level (metres below present)

140 120 100 80 60 40 20 0

Rangitoto

Mt Wellington
Pukeiti, Ōtuataua
Motukorea
Mt Eden, Three Kings
*Flare up
Mt Victoria
Mt Cambria, McLennan Hills
Māngere,
McLaughlins Mt
One Tree Hill

Mt St John

North Head,
Maungataketake

Domain, Grafton
Mt Roskill

Ōrākei Basin, Mt Albert

Boggust Park
Pukewairiki
Albert Park

*Puketūtū
Taylors Hill
Mt Richmond
Crater Hill   Glover Park
Kohuora
Mt Hobson

Tank Farm
Onepoto
Pupuke

Thousands of years ago

0
20
40
60
80
100
120
140
160
180
200

Waitematā and Manukau harbours/forested valleys

Waitematā and Manukau harbours present

↻ Time line for the past 200,000 years showing the time of eruption of Auckland's volcanoes and the variation in sea level.

changed in composition from the mixed broad-leaf–podocarp forest of warm periods, like today, to beech-dominated forest of the coldest periods. The main impact on Auckland and its volcanoes from these climate cycles was the changing sea level. During the warm periods, sea level was close to or just above that of today. During the peaks of the two glacial periods some of the Earth's water was trapped on land in the polar ice caps and, as a result, sea level fell to 120–130 m lower than today. At these times the Waitematā and Manukau harbours were forested river valleys and the coastline had receded to out beyond Great Barrier Island (Aotea) on the east coast and many kilometres off the present west coast.

For most of the time when the volcanoes were erupting, sea level was more than 20 m

lower than today and all the volcanoes, except Rangitoto, erupted on land and not in the sea. Two of the three island volcanoes of today – Motukorea and Puketūtū – erupted on land and subsequently became islands when sea level rose as the world warmed up after the peak of the Last Glacial period (20,000–18,000 years ago). The two coldest (glacial) periods lasted for several thousand years each, ~140,000 and 20,000 years ago, and then the world rapidly warmed and sea level rose over a period of about 10,000 years to reach the two warmest (interglacial) periods. These warm interglacial periods existed for about 10,000 years – the Last Interglacial (~130,000–120,000 years ago) and the Present Interglacial (the last ~10,000 years).

Auckland's next eruption?

The magma below Auckland

Naturally, all Aucklanders want to know whether another volcano will erupt in the city and, if so, where and when, and will we get any warning. To try and answer these questions we have to understand where the magma is coming from and what is triggering small batches of it to rise periodically to the surface and erupt. This requires several kinds of geological study.

Specialised geologists called petrologists cut thin slices of rock and study the composition of Auckland's volcanic basalt with a transmitted-light microscope. Through this they see a closely knit mix of small and very small crystals of several minerals – plagioclase, olivine, clinopyroxene and ores. The larger crystals formed within the magma as it started to cool while rising towards the surface. The smaller crystals grew quickly as the basaltic lava cooled and solidified after it was erupted. The margins of flows and frothy scoria cool so rapidly that there is no time for even small crystals to grow and dense, black glass is formed.

Geochemists, like the University of Auckland's Ian Smith, grind up small samples of basalt or select out single crystals and use expensive modern equipment to study in detail the chemical composition of Auckland's volcanic rocks. They have found that there is a considerable amount of variation in their make-up, with their silica composition ranging between 38 and 50 per cent, which defines them as basalts. Determining the chemistry of these rocks provides considerable insight into the temperature and pressure, and therefore depth, at which the original magma was formed and also information on how the chemistry has changed during the magma's slow or fast ascent.

Some magma batches have risen more slowly or stalled for a short while on their way towards the surface and, therefore, cooled sufficiently

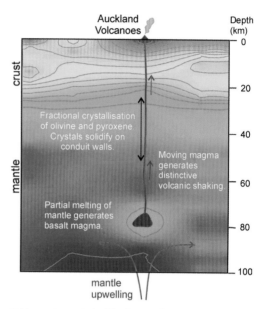

◑ Magma beneath Auckland comes from partially melted mantle rock at about 70–90 km depth. This magma rises to the surface in small batches that usually erupt in a different place each time to form a new volcano. During its ascent to the surface the chemistry of the primitive magma may evolve by the growth within it of minerals that crystallise at higher temperatures and preferentially remove some elements from the remaining molten mix. The background in this diagram is the coloured seismic shear-wave velocity profile kindly provided by geophysicist Nick Horspool.

○ A thin section of basalt rock from a Mt Wellington lava flow as seen through a microscope. The scale is 1 mm long. The larger crystals are of olivine and the brown-tinged, more irregular crystals are clinopyroxene. The crystals forming the matrix are plagioclase (white, stick-shaped), ores (black specks), olivine and clinopyroxene. *Photo by Lucy McGee*

before eruption to temperatures low enough for crystals of some minerals (olivine and clinopyroxene) to start growing inside the melt. The growth of these minerals preferentially removes from the magma the elements that they are made of and this changes the chemistry of the remaining melt. Often these crystals grow on the walls of the conduit at depth and are not erupted. Other times, some of these crystals, 1–10 mm across, are erupted with the molten lava and captured in the basalt as it solidifies. Small, yellow-green crystals of olivine are sometimes visible within the solid basalt or scoria deposits from Auckland's volcanoes.

Geochemical studies show that each batch of magma that rose to the surface and erupted in the Auckland Volcanic Field had a slightly different composition, and sometimes that composition had changed (usually becoming more evolved) during the time it was erupting – a period of weeks to months. Studies show that the primitive parent magmas were formed by partial melting of solid mantle rock called peridotite, at temperatures around 1450 °C.

These temperatures occur at depths of 70–90 km beneath Auckland. Using seismological methods, GNS Science geophysicist Nick Horspool found an area of low velocity for seismic shear waves (indicating less solid mantle rock) at 70–90 km deep and this is inferred to be the magma-producing region for Auckland's volcanoes.

Almost all of Auckland's volcanoes erupted alkali basalt. This has 45–49 per cent silica and formed at about 1450 °C by partial melting of about 2–4 per cent of the parent mantle peridotite. University of Auckland graduate student Andrew Needham found that the lava flows erupted from Rangitoto were subalkali to tholeiitic basalt – highly unusual for Auckland. They contain 49–50 per cent silica and formed at about 1400 °C by partial melting of about 6 per cent of the peridotite at the shallower depth of about 65 km.

These geochemical studies indicate that Auckland's magma originates in the upper mantle, which is quite different from hot-spot mantle plume volcanoes, such as Hawai'i, where the magma originates much deeper. The questions now being asked are, why is this partial melting present beneath Auckland and why do small batches sometimes rise to the surface to produce the Auckland volcanoes above? While there is no obvious direct connection between Auckland's volcanoes and the boundary between the Australian and Pacific tectonic plates, their close proximity to that boundary may not be entirely a coincidence. One hypothesis is that this area is under tension, being slowly pulled apart at depth, and that maybe there is some upwelling of hotter mobile mantle from beneath, which slightly increases the temperature and reduces the pressure in this zone of partial melting and magma generation.

⊙ Computer graphic of a future eruption in Auckland. In this instance fire-fountaining is portrayed erupting near Māngere Bridge, but in reality we do not know where the next eruption will break out. *Courtesy of Ministry of Civil Defence and Emergency Management*

Time and place of Auckland's next eruption

The Auckland Volcanic Field has been erupting on and off for the last 200,000 or so years. The most recent eruption was Rangitoto just 620–600 years ago and thus it is quite reasonable to expect that there will be more eruptions in the future, but where, when and of what size we do not know.

Where?

Each of Auckland's existing volcanoes has erupted only once. After eruptions at each stopped, magma that was in the plumbing immediately beneath the volcano cooled and solidified and appears to have blocked the conduit forever. Next time magma rises from the deep and approaches within a few kilometres of the surface it will encounter the blocked plumbing of previous volcanoes and will find an easier route, probably up a different section of fault line, before erupting as a new volcano. Thus it is unlikely that the next eruption will be a reawakening of any of Auckland's existing 'extinct' volcanoes but instead will be somewhere where no volcano has previously erupted. We can also predict that the next eruption is likely to be within or just outside the boundaries of the existing volcanic field – somewhere between Takapuna in the north and Wiri in the south and between Mt Albert in the west and Pakuranga in the east.

Some people have speculated that the next eruption will be in the northeastern part of the field, possibly in the Waitematā Harbour. This speculation is based on the fact that the two youngest eruptions – Rangitoto and Mt Wellington – have been in this sector. However, the theory is not well supported by our present knowledge, which appears to show no pattern or trend in the distribution of volcanoes of different ages. For example, both the oldest (Pupuke and Onepoto) and the youngest (Rangitoto) volcanoes lie in the northernmost sector of the field, and a wide range of volcanoes of different ages are also present in the central and southern sectors.

When?

At the present time we have no way of predicting when the next eruption might occur. It could be next month or not for thousands of years. When the first signs of moving magma at depth (~20 km down) are detected we may have only a few days' to a few weeks' warning of a pending eruption in the city.

Now that we know with moderate accuracy the ages of most of Auckland's 53 volcanoes, we can state categorically that there is no predictable pattern in the time between successive eruptions. Twelve of Auckland's volcanoes are older than 100,000 years and 10 more erupted between 100,000 and 40,000 years ago. Half of the volcanoes (26) erupted in the 21,000-year interval between 40,000 and 19,000 years ago, with only five volcanoes erupting since then.

The age distribution of volcanoes indicates a gap of about 10,000 years between Rangitoto and the second- and third-youngest eruptions (Purchas Hill and Mt Wellington) and another 5000-odd years back to the eruption prior to that. Does this indicate that Auckland's next eruption is likely to be at least a few thousand years away? We do not know. It is also possible that the next eruptions could come in a bunch, like they did 30,000 or so years ago.

How big?

The size and nature of the next eruption is also unknown. By far the largest batch of magma to erupt was from the youngest volcano – Rangitoto. Does this indicate that the nature of Auckland's volcanoes is changing towards more voluminous eruptions? There is little reason to think so but it cannot be ruled out. Many smaller volcanoes have been produced in the time between the eruptions of Rangitoto and the second-largest volcano, One Tree Hill, 60,000 years ago. Currently, we have no way of knowing how large the next volcano will be and, like all the other volcanoes, the style of its eruption will depend on the amount of water present where it breaches the surface and the volume of magma that comes up.

○ View from the south over three of the oldest volcanoes in the Auckland Volcanic Field – Onepoto, Tank Farm and Pupuke on the North Shore. The north–south alignment of all three suggests that the magma that fed each came up the same linear fault or zone of weak rock.

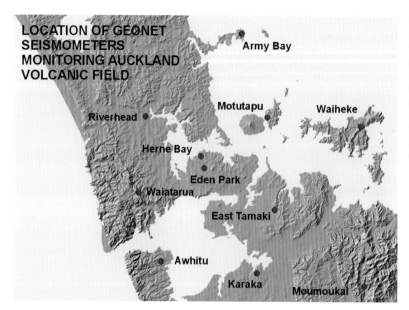

LOCATION OF GEONET SEISMOMETERS MONITORING AUCKLAND VOLCANIC FIELD

Army Bay

Motutapu

Waiheke

Riverhead

Herne Bay

Eden Park

Waiatarua

East Tamaki

Awhitu

Karaka

Moumoukai

◑ Map showing the location of the 11 permanent seismometers that are constantly monitoring the Earth beneath the Auckland Volcanic Field for any sign of moving magma at depth that might signal an imminent eruption in the city.

Monitoring for volcanic activity

The only method we have available at present for determining the time and place of the next eruption in Auckland is to detect and track moving magma deep beneath the city. The Earthquake Commission and GNS Science's GeoNet Project is currently using a network of 11 seismometers to monitor the crustal region above the source area for the magma, some 70–90 km beneath the surface. As the magma moves up and through the brittle crust (0–20 km depth) it should generate vibrations that travel through the ground and can be detected by these sensitive motion detectors. Seismometers also detect normal earthquakes, but the seismic signature of moving magma as it gets closer to the surface is distinct from the seismic waves generated by earthquakes.

The seismometers have been installed on or in solid ground well away from main roads and other sources of background seismic noise. Half the seismometers are installed hundreds of metres below ground in boreholes so as to lessen the surface vibrations of traffic and other activities. Each seismometer measures

the precise arrival time and strength of seismic waves and these will be used to more accurately determine the location of the moving magma and its progress towards the surface. The data from the seismometers is transmitted to GeoNet as it is collected. The data is then monitored by specialised computer software that is able to automatically detect locally generated seismicity. If this happens, a duty scientist will examine the record as soon as it is received (at any time of day or night) and aim to determine whether the earthquakes are likely to be caused by moving magma or not. With this monitoring system, Auckland Emergency Management can be notified within an hour that there could be a pending volcanic eruption in Auckland.

The seismometers will detect the faint signals of deep magma movement well before it is ever felt by people on the surface. Overseas studies suggest that there should be two days' to two weeks' warning between the time of initial detection (perhaps at about 20 km depth) and eruption at the surface. Early on, while the magma is still down deep, it will not be possible to determine in which part of the field the new volcano will erupt, but as the magma approaches the surface the location will be able to be predicted with increasing accuracy. It is not expected that the detection of moving magma at depth will always lead to an eventual eruption. On some occasions the moving magma may stop, never to move again, or stop temporarily and then move on days, weeks or months later.

Auckland Emergency Management and other agencies have prepared plans on how they will manage a pending volcanic eruption in Auckland with evacuation of several hundred thousand people from the threatened parts of the city. Only the suburbs close to the impending eruption (within ~5 km) will need to be evacuated. It is clear that news of the seismic record of the moving magma will become widely known and communicated to the public and that many people will voluntarily leave the city even before an evacuation is ordered. Equally, there will also be residents who will refuse to leave even after an evacuation has been ordered.

◉ An earthquake analyst at GNS Science's GeoNet Project examines an earthquake waveform like those that will give warning of the next eruption of Auckland's Volcanic Field. *Courtesy of New Zealand GeoNet Project*
◉ The most dangerous phase of the next eruption in Auckland is likely to be an explosive pyroclastic base surge of searing gas and ash, which might smother and kill everything in its path for up to 5 km from the vent. This photo, from the Mt Pinatubo eruption in 1991, illustrates the pyroclastic surge style of eruption. *Courtesy of US Geological Survey*

The next eruption – what to expect

As rising magma approaches the surface and an eruption is imminent, the intensity of earthquake shaking will increase and buildings in nearby suburbs could be damaged. In the hours, perhaps days, before the eruption commences, the ground directly above the magma may start to bulge upwards and crack. By then everybody should have evacuated this part of the city, as the initial phases of an Auckland volcanic eruption are potentially the most dangerous and destructive.

If the rising magma encounters ground or surface water – and in the majority of Auckland eruptions this seems to have been the case – then the initial eruptions will likely be of the wet explosive style. If the next eruption is located in deeper parts of the Waitematā or Manukau harbours, there is a possibility that the initial outburst will generate a small tsunami that could sweep ashore and cause damage in low-lying coastal suburbs, such as Onehunga, Māngere Bridge, Mission Bay, Kohimarama, St Heliers, Devonport and Cheltenham, as well as the port of Auckland or Auckland Airport.

In explosive eruptions, all traces of suburbia will be destroyed in the area surrounding the large crater that is formed. These craters usually have a diameter of 500–1000 m and, beyond this, the most dangerous and damaging parts of explosive eruptions are the base surges of searing gas and ash. They explode out in all directions from the vent and may devastate an area 1–5 km from the eruption's centre. These explosive surges are usually preceded by powerful shock waves strong enough to knock down trees and break windows and these impacts may extend beyond those of the base surge itself. The rapidly travelling super-heated blast arrives a few seconds later and will kill all humans and animals in its path and destroy all vegetation and houses. Only the skeletons of the more strongly constructed buildings may remain. Base surges will travel further over flat land or sea than over more hilly suburbs. The base surges and ash fall will build a low tuff ring around the crater and will bury any structures still standing after the initial base surge blasts.

The wet explosive style of eruption also blasts volcanic ash a few kilometres into the air. The height, volume and duration of these ash eruptions, as well as the strength and direction

◔ In 1973, fiery explosive and fire-fountaining eruptions, similar to those that built Auckland's volcanic cones, buried houses with scoria in Heimaey, Iceland. The same scenario may happen in the latter phases of Auckland's next eruption. *Courtesy of US Geological Survey*

◑ In the latter phases of Auckland's next eruption, molten lava is likely to flow down valleys and streets progressively burning and destroying everything in its path, as here in Hawai'i. *Courtesy of US Geological Survey*

of the wind, will determine the extent of damage and disruption they will cause. Much of the ash will fall back to the ground and contribute to the construction of the tuff ring close to the crater, but wind may blow a blanket of ash a few centimetres to tens of centimetres thick over 10 km or more from the vent into areas not evacuated. In areas of thick ash fall, darkness may descend and people without respirators will experience breathing difficulties. Ash is abrasive and if it gets into motors it can cause damage to machinery and vehicle breakdowns. If the airport is not within the evacuation zone, it is likely to be closed as a precaution against ash damage to aircraft. Ash will build up on roads, causing further major traffic congestion.

Ash will also accumulate on house roofs and if it is wet, or some tens of centimetres thick, its weight could cause many to collapse. If heavy rain accompanies or follows the ash eruptions, there is a likelihood that many stormwater drains in downwind suburbs will become blocked and widespread flooding may occur. Wet ash, erupted wet or falling with rain, may stick to trees, power lines, insulators and cell phone towers, causing branches to

fall, potential electricity flashovers with power outages and communications disruption in parts or all of the city. Ash fall will contaminate water, often making it less palatable rather than toxic, and, given the right wind direction, could temporarily close freshwater supply to the city from either the Waitākere or Hunua catchments. In some areas power outage could impact essential services, such as pumping stations for freshwater supply and sewerage reticulation.

After the explosive eruptions are replaced by dry eruption styles of fire-fountaining, fiery explosions and lava flows, the hazards to life and property will greatly diminish and be more predictable. However, a return to explosive eruptions could not be ruled out, creating a difficult emergency management environment. The hazards from these dry styles of eruption will be mostly limited to within the area that may have already been devastated by the explosive eruptions and base surges. During fire-fountaining, scoria descends close to the vent and builds up a scoria cone, and fine scoriaceous ash may be blown a few kilometres away. Wind-blown incandescent scoria can start fires, although boarding up windows may reduce this risk to houses.

Periodically, volcanic bombs, up to a metre across, may be blasted high into the air and land within a kilometre of the vent, but these would usually only be a threat to volcanologists who may be allowed this close to take samples for monitoring and predicting the course of further eruptions.

If the supply of magma is sufficient, lava flows may pour from around the base of the rising scoria cone and spread over low-lying land or flow down valleys. The speed of these flows would usually allow all pets and many moveable possessions to be evacuated, but any vegetation or buildings in the way are likely to be set on fire and probably demolished and consumed by a lava flow. In some instances the speed of flows can be slowed or the lava halted by hosing cold water on them. Bulldozers can build barriers that may divert the flows but this could be difficult in built-up suburban areas. Small secondary explosive blasts may occur as lava flows meet cold-water bodies, such as streams, ponds or swimming pools.

Volcanic ash clean-up is time-consuming and costly, and disposal of ash requires substantial land and planning. Ash that is not removed can be remobilised by wind as dust storms for years or even decades after the initial ash falls. Also, a reactivation of eruptions could occur after a period of quiet, complicating emergency man-agement, particularly if clean-up and population return is under way. If people and businesses are evacuated for more than months, those perma-nently leaving the city will steadily increase.

In summary, the next eruption in Auckland will cause considerable economic and social disruption to both the city and New Zealand, and has the potential to completely destroy several suburbs. If people evacuate the target zone when instructed to do so by Auckland Emergency Management, then they, most of their pets and a few of their most treasured belongings will survive the volcanic activity. During the eruption, life in the remainder of the city will likely be disrupted by traffic congestion, power outages, the potential impacts of ash fall and unexpected house guests from the evacuated area.

DEVORA

DEVORA is an acronym of **De**termining **Vo**lcanic **R**isk in **A**uckland. It is a multi-agency, trans-disciplinary collaborative research programme led by volcanologists at the University of Auckland and GNS Science. It is funded by the Earthquake Commission and Auckland Council. The project began in 2008 and is aimed at improving our understanding of the volcanic hazard in Auckland and assisting with formulating plans to deal with it. Scientists, emergency managers, economists and other stakeholders are working together to create an integrated risk model summarising the answers to three major questions:

1. How, why, how often and how fast does magma move to the surface in the Auckland Volcanic Field?
2. What might happen when the next volcano erupts in Auckland?
3. What are the potential impacts on the city and citizens?

Researchers are working with Auckland Emergency Management to incorporate their findings into policy, and with lifelines organisations and businesses to improve their resilience to volcanic disasters.

Auckland Lifelines Group

Auckland Lifelines Group, a voluntary consortium of Auckland local government, Auckland Emergency Management, public utility providers, engineers and scientists, has been investigating the likely impacts of future Auckland eruptions on essential energy, communication and transport networks. It has been developing strategies and designs for important infrastructure that will lessen the impact not only of an eruption from within Auckland but also one from further afield.

Human interaction with Auckland's volcanoes

Māori occupation and use of Auckland's volcanoes

In pre-European Māori times (1350–1840), the Auckland Volcanic Field became one of the most densely populated areas in Aotearoa New Zealand. This happened for a number of reasons. Firstly, because of the location of Tāmaki Makaurau/Auckland on the narrowest isthmus between the great sheltered waterways of the Hauraki Gulf, the Manukau Harbour and the Waikato River. It was strategically and economically important to control the portages between these eastern and western waterways. Another reason was the availability of rich food resources from the Manukau and Waitematā harbours. A third reason was the extensive areas of rich volcanic soils suitable for growing crops.

Gardening

Traditional and archaeological evidence indicates that the rich volcanic soils and cones of the Auckland Volcanic Field were highly attractive to early Māori colonisers and were widely occupied by about 650 years ago, and remained so up until the arrival of Europeans in the 19th century. The soils of the volcanic ash-mantled land and lava-flow 'stonefields' were more fertile than the surrounding clay soils derived from older sedimentary rocks. They were friable and relatively easy to cultivate using hand-held wooden implements. They were also warmer

◔ Pre-European Māori gardens and settlements on the Matukuturua lava-flow fields on the south side of Matukutūreia/ McLaughlins Mountain, Wiri. *Painting by Chris Gaskin*

35

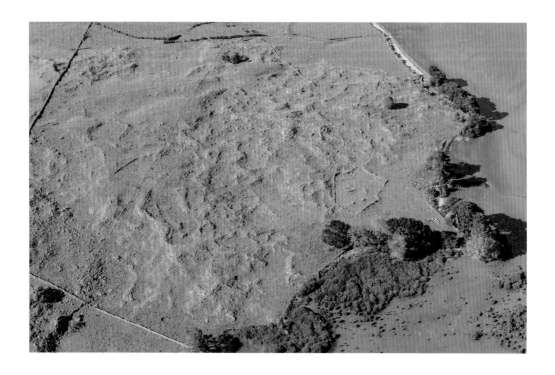

⬆ Stone walls and mounds that were part of an extensive pre-European gardening system established on the lava flows from Ōtuataua Volcano, Māngere. *Photo by Alastair Jamieson*

⬇ Sketched cross-section showing how pre-European archaeological features on the scoria cone pā were created out of the original slopes, with wooden structures that once sat above them.

steep banks
cut and built up
for defence

terrace built
with fill

pit remains as
lower part of
food storage
shed

ditch dug
for defence

original cone slope

posthole impressions
from house supports

row of posthole
impressions from
defensive pallisade

midden below terrace

throughout the year because of their stony nature, providing a longer growing season for the subtropical plants introduced from the Pacific by Māori, such as kūmara, taro, yams and gourds. The earliest Māori arrivals might have been able to survive as hunters and gatherers of the natural resources but, as the population built up, larger prey species, such as seals and moa, disappeared and cultivations became an increasingly important food source.

On lava-flow stonefield gardens a large amount of surface rock had to be cleared and heaped up. The heaps were often used as raised rock-and-earth garden mounds, which increased soil moisture and temperature, providing a longer growing season for frost-tender crops and nursery beds to get kūmara off to an early start prior to being planted out in the main gardens. The cleared rocks were also used to form low boundary walls that often extended out from the cones like the spokes of a wheel between gardens of different family groups. These areas were subdivided by smaller stone walls and stone alignments for different crops. Some stone alignments also formed the base of brush windbreaks that sheltered the gardens, while others demarcated pathways. Rectangular and C-shaped stone alignments indicate the sites of cooking shelters and houses.

Most of the pre-European stonefield garden features of Auckland were modified by 19th- and 20th-century European pastoral farming practices and have now been destroyed by urban sprawl. Small areas of less-modified stonefield garden are protected within Ōtuataua Stonefields Historic Reserve, Matukuturua Stonefields

Historic Reserve, Wiri Stonefields Reserve and Maungarei Stonefields Reserve.

Settlement and defence

Māori, who were attracted to the rich volcanic soils of the lava-flow fields, initially used the associated scoria cones for settlements. The smaller cones were occupied by as many as several hundred people living in scattered hamlets located on the cone's slopes. The largest cones were sprawling settlements occupied by much larger groups of related families. The main archaeological modifications of the cones are the hand-dug terraces on the slopes, crater rims and high points. The terraces provided

◑ Most of the scoria cones in the Auckland Volcanic Field were used as defensive pā by pre-European Māori, as depicted in this painting of Māngere Mountain. Roof-covered pits were used for storing kūmara; ditches, banks and wooden stockades were used for defence. *Painting by Chris Gaskin*

Terracing and pits dominate this view of the slopes of Te Pane-o-Mataaho/Māngere Mountain and provide archaeological evidence of the extensive occupation of this important cone in pre-European times. *Photo by Alastair Jamieson*

Maungakiekie/One Tree Hill is the most extensively terraced of all of Auckland's volcanic cones and one of the largest pre-European archaeological site complexes in New Zealand. It includes dozens of house sites and garden terraces, and numerous groups of food-storage pits. This scoria cone pā is one of the largest pre-Iron Age forts in the world. Its four summits were all heavily defended by ditches and wooden palisades. On its peak was the tihi, the most sacred and heavily defended part of the complex.

flat areas for the construction of houses, cooking shelters and kūmara pits. These food-storage pits are usually rectangular in plan and several metres deep. They were covered by steeply pitched roofs made from tree fern trunks covered with dirt. These 'underground' buildings were surrounded by drains, secure from pests and accessible through small, tightly closing wooden doors. They were dry, warm and of uniform humidity – well suited for the long-term storage of kūmara and other delicate subtropical food crops.

Fortification of the cones to become pā was not really needed until the 16th century when population pressure and increasing tribal separation led to feuds and local warfare. The rich volcanic soils were sought after by those who did not have them, and the harvested kūmara in their storehouses needed protection from marauding thieves. More intensive earthwork fortifications emerged in the 17th century when inter-regional conflict intensified. To fortify the cones, the slopes were artificially steepened, particularly at the front and back of terraces, wooden stockades were erected along the edges of some of the terraces, and ditch and bank defences were constructed across ridges and spurs that otherwise created easier routes for attack. Although the wooden portions of these pā and buildings have long since disappeared, the earthworks associated with them still remain on almost all of the scoria cones in Auckland, reminding us of their central role in this period of the human history of our country.

Volcanoes as water sources

Fresh water is a basic need of all humans. Throughout the many centuries of pre-European Māori settlement, permanent sources of fresh water were of critical importance to their survival. Within the Auckland Volcanic Field there are few surface streams because rainwater rapidly soaks into the porous volcanic rock and soil, often flowing down the old stream valleys of the pre-volcanic landscape to emerge, often near the coast, as springs. Thus, the location of permanent freshwater sources, such as springs, determined the ability of Māori to occupy the volcanic field, in particular during long, dry summers. The springs generally had to be within reasonable walking distance of permanent kāinga (settlements) and, therefore, had a direct influence on settlement patterns. All these freshwater bodies were named and treasured, with many commemorating ancestors who were directly associated with them, or important historical events.

The European town of Auckland was founded in 1840, and it grew quickly, centred around Queen St. Further afield, additional small independent townships were progressively established in places like Onehunga, Panmure, Howick, Ōtāhuhu, Devonport and Birkenhead. For the first 26 years, Auckland depended almost entirely for its freshwater supply on rainwater collected in house tanks and on a few indifferent wells. This was because the Waitematā Sandstone rocks that underlie the Auckland Isthmus have

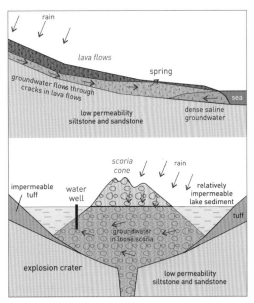

● Upper: Cross-section showing groundwater flowing downslope through the fractures in a basaltic lava flow and emerging at the surface as springs when it encounters dense, underground salt water at sea level.
Lower: Cross-section showing the groundwater resource within loosely packed scoria trapped inside the impervious walls of an explosion crater.

poor water-bearing capacity. This was a period of chronic water shortages, with queues at the few producing public wells and a lack of water to fight major fires in 1858 and 1863.

Colonists dug exploratory water wells all over the isthmus and discovered that the best sources of groundwater were in the highly permeable scoria deposits or fractured lava flows of Auckland's volcanoes. The ash-lined volcanic explosion craters were largely impermeable basins that collected rainwater. In most instances, except Lake Pupuke, the craters were filled with sediment up to their overflow points. Where this fill is mostly mud (e.g. the tidal lagoons), there is little room left for groundwater, but where the fill is loose scoria or highly fractured basalt of ponded lava flows or a former lava lake, the deposit is fully charged with water.

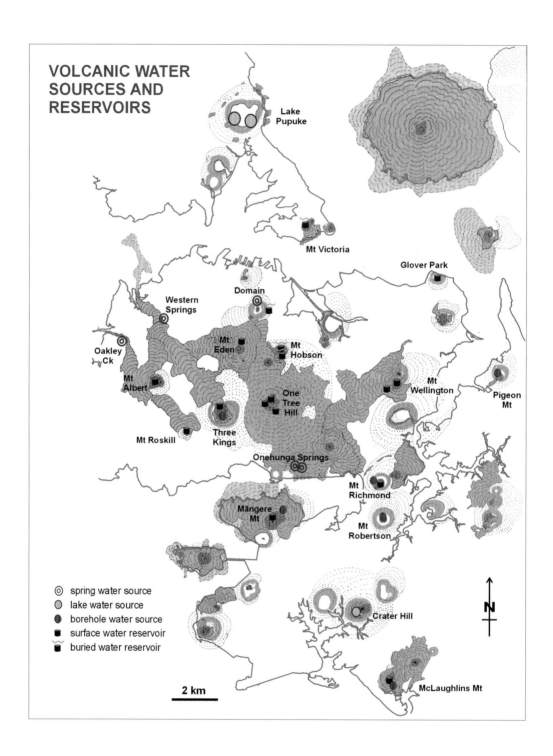

VOLCANIC WATER
SOURCES AND
RESERVOIRS

Lake
Pupuke

Mt Victoria

Glover Park

Domain

Western
Springs

Mt
Eden

Mt
Hobson

Mt
Wellington

Oakley
Ck

Mt
Albert

One
Tree
Hill

Pigeon
Mt

Mt Roskill

Three
Kings

Onehunga Springs

Mt
Richmond

Māngere
Mt

Mt
Robertson

◎ spring water source
◯ lake water source
● borehole water source
■ surface water reservoir
■ buried water reservoir

Crater Hill

2 km

N

McLaughlins Mt

The two largest volcanic-related sources of fresh water for the growing city of Auckland were Lake Pupuke on the North Shore and Western Springs. Here is the Western Springs pumphouse in 1880 (now a centrepiece of MOTAT), which pumped water that poured out of cracks in the lava flows from Three Kings up to reservoirs on the high ridges above the central city between 1877 and the mid-20th century. Conical cone of unquarried Mt Albert in the distance. *Photographer unknown, Auckland Museum*
Map showing the sources of water from Auckland's volcanoes that were tapped to supply early Auckland City. The locations of water reservoirs on Auckland's volcanic cones are also shown.

Auckland's lava-flow basalts are highly fractured by cooling cracks and mixed with zones of broken and scoriaceous rock that make excellent underground aquifers. Many of these lava flows swept down existing stream valleys and, today, much of the water that originally flowed down these streams now flows through the solidified basaltic lava rock to resurface as springs as it approaches the coast and meets saline groundwater infiltrating from the sea.

In the early days of Auckland, these volcanic-related water sources attracted considerable use, for watering horses and for industries that required fresh water for processing, to drive waterwheels or supply steam engines. Flour mills, for example, were located near the Domain ponds (1840s), at Western Springs (1850s), Oakley Creek (1850s) and Bycroft Springs, Onehunga (1850s on). Breweries were established around Seccombe's Well in a Mt Eden lava flow in Khyber Pass Rd from the 1850s on; and abattoirs at Western Springs (1870s) and later Westfield, Mt Richmond.

41

Volcanoes lost and damaged

When Auckland City was founded in 1840, 38 volcanoes had intact scoria cones, albeit slightly modified on the surface by centuries of Māori occupation. As of now (2019), 15 of these volcanoes have had their cones completely removed by quarrying and a further nine have been fiercely ravaged. Only two have cones essentially untouched by quarrying – Rangitoto and Motukorea.

In the city's early days, the scoria cones were the main source of road materials. Scoria was cheap to extract as it was unwelded and loose, and could be shovelled straight from the quarry face onto the back of a wagon. By the late 20th century scoria had been largely replaced by crushed basalt from the lava flows or greywacke rock from the Hunua Ranges, although smaller quantities of scoria continued to be used for free-draining fill.

Early on, many of the volcanic cones were set aside as Crown-owned public domains for recreation or quarry reserves. There was no requirement for the local domain boards to protect their volcano's heritage values. The only income for their upkeep or development came from grazing leases. Many boards made ends meet by selling off the quarrying rights, often by decision of the local borough council. Examples included Mt Cambria, Three Kings,

Little Rangitoto, Mt Wellington, Crater Hill, Pigeon Mountain, Green Mount, Ōtuataua and Māngere Mountain. New Zealand Railways quarried away the upper third of Mt Albert, all of Mt Smart and most of Wiri Mountain for ballast beneath their rail lines.

When humans arrived, 19 of Auckland's volcanoes had essentially complete explosion craters surrounded by tuff rings. None have escaped some form of modification. The least modified are Pupuke, which still contains its original freshwater lake, and Tank Farm at Northcote, which still has its intertidal mangrove forest and fringing salt marsh. Four of the others (Grafton, Styaks Swamp, Ash Hill, Cemetery Crater) are completely gone – hidden beneath industrial and residential subdivisions. Ōrākei Basin and Te Hopua have been modified by the construction of a major railway embankment or motorway through the heart of their craters. Panmure Basin, Crater Hill and Pukewairiki all have major roads cutting through one side of their respective tuff rings, and the restored Māngere Lagoon has an artificial causeway for a major sewer line built across the western side. Three of the seven explosion craters that were tidal inlets when humans arrived have been 'reclaimed' and now have dry crater floors – Onepoto, Te Hopua and Pūkaki Lagoon.

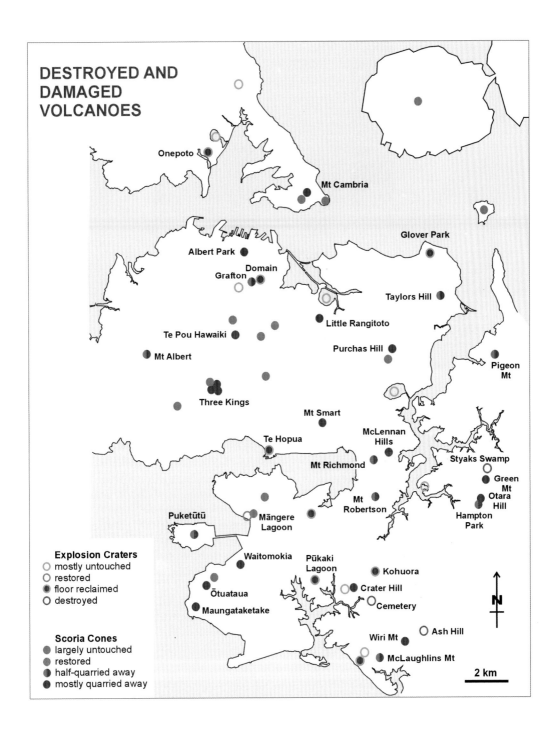

DESTROYED AND DAMAGED VOLCANOES

Onepoto

Mt Cambria

Glover Park

Albert Park

Domain

Grafton

Taylors Hill

Little Rangitoto

Te Pou Hawaiki

Purchas Hill

Mt Albert

Pigeon Mt

Three Kings

Mt Smart

Te Hopua

McLennan Hills

Styaks Swamp

Mt Richmond

Green Mt

Puketūtū

Mt Robertson

Otara Hill

Māngere Lagoon

Hampton Park

Waitomokia

Pūkaki Lagoon

Kohuora

Explosion Craters
○ mostly untouched
○ restored
◑ floor reclaimed
○ destroyed

Ōtuataua

Crater Hill

Maungataketake

Cemetery

Scoria Cones
● largely untouched
● restored
◑ half-quarried away
● mostly quarried away

Ash Hill

Wiri Mt

McLaughlins Mt

N

2 km

Tūpuna Maunga Authority

In July 2014, as part of a Treaty of Waitangi claims settlement, 14 Crown-owned cones (Tūpuna Maunga) in the Auckland Volcanic Field were vested with Ngā Mana Whenua o Tāmaki Makaurau, on the basis that they are held in trust for the common benefit of the iwi/hapū of Ngā Mana Whenua o Tāmaki Makaurau and the other people of Auckland. The Tūpuna Maunga are collectively owned by the 13 iwi/hapū of Ngā Mana Whenua o Tāmaki Makaurau via the Tūpuna Taonga o Tāmaki Makaurau Trust. The 14 maunga (volcanoes) are: Matukutūruru/Wiri Mountain, Maungakiekie/One Tree Hill, Maungarei/Mt Wellington, Maungauika/North Head, Maungawhau/Mt Eden, Ōhinerau/Mt Hobson, Ōhuiarangi/Pigeon Mountain, Ōtāhuhu/Mt Richmond, Ōwairaka/Te Ahi-kā-a-Rakataura/Mt Albert, Puketāpapa/Pukewīwī/Mt Roskill, Rarotonga/Mt Smart, Takarunga/Mt Victoria, Te Kōpuke/Tītīkōpuke/Mt St John and Te Tātua-a-Riukiuta/Big King.

The Tūpuna Maunga Authority now administers these reserves except for Rarotonga/Mt Smart, which is currently managed by Auckland Council as Mt Smart Stadium. Another maunga, Te Ara Pueru/Te Pane-o-Mataaho/Māngere Mountain, is also administered by the Tūpuna Maunga Authority although it is retained in Crown ownership.

The authority consists of 13 members – six appointed by Ngā Mana Whenua o Tāmaki Makaurau, six appointed by Auckland Council and one non-voting member appointed by the Minister for Arts, Culture and Heritage. Auckland Council is responsible for the routine management of the maunga under the direction of the independent Tūpuna Maunga Authority. The new authority has started implementing some of their new management philosophy of removing vehicles, grazing stock and exotic trees from some of the maunga.

⊙ View from the north across the craters and summit of Maungakiekie/One Tree Hill scoria cone. This is one of the mauna now administered by the Tūpuna Maunga Authority. *Photo by Alastair Jamieson*

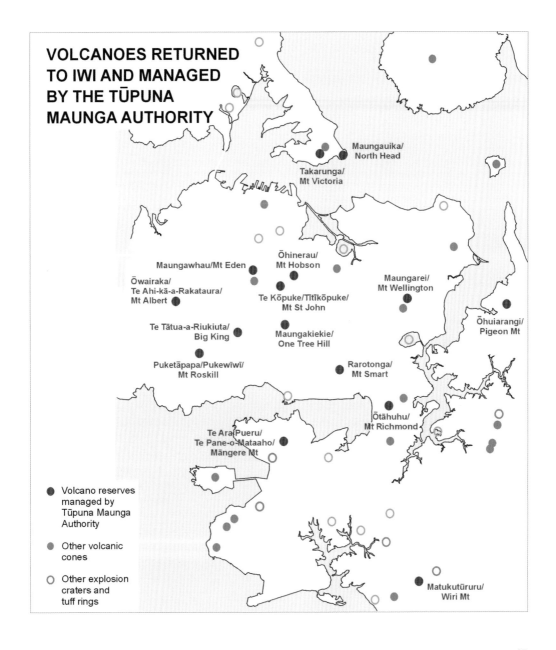

VOLCANOES RETURNED TO IWI AND MANAGED BY THE TŪPUNA MAUNGA AUTHORITY

Maungauika/
North Head

Takarunga/
Mt Victoria

Ōhinerau/
Mt Hobson

Maungawhau/Mt Eden

Ōwairaka/
Te Ahi-kā-a-Rakataura/
Mt Albert

Maungarei/
Mt Wellington

Te Kōpuke/Tītīkōpuke/
Mt St John

Te Tātua-a-Riukiuta/
Big King

Ōhuiarangi/
Pigeon Mt

Maungakiekie/
One Tree Hill

Puketāpapa/Pukewīwī/
Mt Roskill

Rarotonga/
Mt Smart

Ōtāhuhu/
Mt Richmond

Te Ara Pueru/
Te Pane-o-Mataaho/
Māngere Mt

● Volcano reserves
managed by
Tūpuna Maunga
Authority

● Other volcanic
cones

○ Other explosion
craters and
tuff rings

Matukutūruru/
Wiri Mt

Volcanoes of the Waitematā Harbour and North Shore

Eight volcanoes have erupted in the northern part of the Auckland Volcanic Field. Two are located in the Waitematā Harbour and six on the North Shore. Rangitoto is the only one of Auckland's volcanoes known to have erupted in the sea; Motukorea, also located in the harbour, appears to have erupted on land, when the sea level was lower. On the North Shore there are three volcanoes at Devonport and three in the Takapuna–Northcote area. All three Devonport volcanoes (North Head, Mt Victoria and Mt Cambria) produced scoria cones and erupted lava flows, whereas the three near Takapuna (Pupuke, Tank Farm and Onepoto) erupted explosively, forming large craters surrounded by tuff rings. In the north we have both the youngest (Rangitoto) and oldest (Pupuke and Onepoto) volcanoes in the Auckland Volcanic Field.

One of the North Shore's volcanoes has been quarried away (Mt Cambria) and one of the intertidal craters has been reclaimed (Onepoto), but the rest are in good shape. All the scoria cones were modified by pre-European earthworks and used as defended pā. The two most prominent cones at the entrance to the inner Waitematā Harbour (North Head and Mt Victoria) were further excavated by Europeans for naval forts. Today the cones or explosion craters of all remaining volcanoes, as well as the stump of Mt Cambria, are reserves.

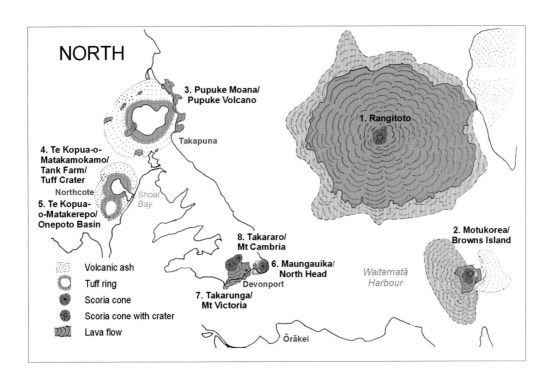

◐ Maungauika/North Head scoria cone (right) with Rangitoto Island shield volcano beyond, 2018.

Rangitoto

Visitors to iconic Rangitoto Island arrive by ferry
at Rangitoto Wharf on the south side of the island
(foreground) and usually walk the track to the summit
for views over Auckland. *Photo by Alastair Jamieson*

◐ Aerial view of the southern slopes of Rangitoto Island showing the different vegetation (darker) growing on the main scoria cone compared with the pōhutukawa forest on the lava flows. In the foreground, branching lobes of some of the youngest lava flows formed as they entered the sea. *Photo by Alastair Jamieson, 2009*

Land status: Rangitoto Island is a Crown-owned scenic reserve, as is adjacent Motutapu Island Recreation Reserve. Both are managed by the Department of Conservation and fully open to the public.

How to get there: Catch a Fullers ferry from the ferry buildings in Quay St, downtown Auckland or from Devonport Wharf. They take passengers to and from Rangitoto Wharf several times a day. See www.fullers.co.nz/destinations/rangitoto-island/ for times and fares. Make sure you know the time of the last returning ferry because if you miss it you may have to hire an expensive water taxi to bring you back.

What to do: Spend a half or full day on Rangitoto Island. On a half-day trip there is time to walk to the summit for a panoramic view of Auckland and the Hauraki Gulf (2 hours return) and appreciate the rubbly lava flows and steeper scoria cone at the top. There may also be time for a quick visit

Some of the lava flows traversed by the track to the summit from Rangitoto Wharf have a surface of broken-up slabs of basalt (up to 1 m across) that probably cooled and solidified on the top of relatively fluid pahoehoe lava. As these flows cooled they became more viscous and slower, and this surface crust was ripped apart and rafted along as slabs.

to the lava caves. An alternative for less fit visitors is to pay for a road-train ride on the gravel road around the island with a stop near the summit for an opportunity to climb the wooden boardwalk and steps to the summit.

On a full-day visit (6 hours) there is time to walk to the summit, visit the lava caves and return to Rangitoto Wharf on the rocky road via Islington Bay (east side) or McKenzie Bay (west side). Both routes are 7 km. Around Islington Bay and Rangitoto wharves there are additional picturesque coastal walks past the sites of many of the former holiday baches that once lined the rocky shore. Just a few baches are still standing as historic reminders of a bygone era. For those interested in plants, examine the plant successions on the lava flows and the scoria cone using the main tracks or from side tracks, such as Kidney Fern and Wilson's Park tracks. On a hot summer's day, swim at the sandy beach at McKenzie Bay (1 hour's fast walk each way from Rangitoto Wharf) or, if the tide is in, there is a small artificial pool near Rangitoto Wharf. There is no shop on the island and the only toilets and fresh water are at Rangitoto Wharf, Islington Bay and McKenzie Bay.

This 2-metre-high, conical tower of basalt spatter near the track to the lava caves is called a hornito. Rangitoto is the only place in New Zealand where hornitos are preserved. Hornitos are small mounds of welded spatter built by the accumulation of incandescent lava ejected through holes in the roof of a lava tube within a lava flow.

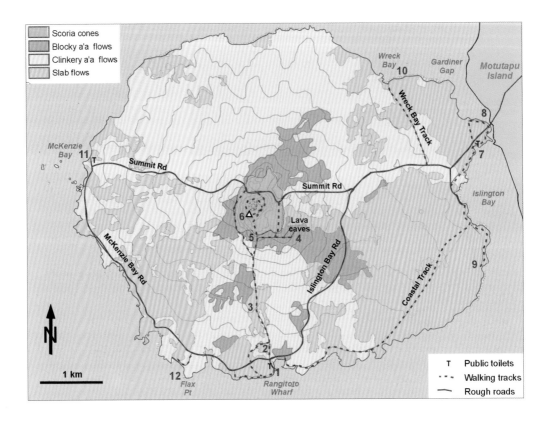

◔ Profile of Rangitoto from Ōkahu Bay showing the gentle lower slopes of the lava-flow shield surmounted by the steeper main scoria cones. The two secondary scoria cones on either side of the main one are clearly visible from this part of Auckland City.

⊙ *Places of interest on Rangitoto Island:*

1. Rangitoto Wharf. Shelter with information panels and water. If the tide is low, look over the sea wall adjacent to the shelter to see a well-exposed branching pahoehoe lava flow that was formed as molten lava reached the sea.
2. Second World War parade ground. This originally was an unfinished tennis court constructed by the prisoners who were housed in a camp close by.
3. A'a and slab pahoehoe lava flows and levees. Look for trenches margined by ridges of rubble (called levees) formed where molten lava drained out of the centre of the lava flow and the overlying rubble collapsed into the empty conduit, producing a trench.
4. Side track to the lava caves (10 minutes each way) leaves the Summit Track at the foot of the scoria cones. Track leads to a 100-metre-long cave with an unusual trench-like shape and near-vertical walls. This was the internal conduit for lava that flowed along within an a'a flow with an outer carapace of solidified clinker blocks of basalt. When the lava drained out it left this long lava tube. Entry and exit points are where the roof has collapsed. If you intend to explore these caves you should take a torch and wear a thick hat for protection in case you bump your head on the roof.
5. Depressed moat where the track to the caves branches from the Summit Track. It separates the top of the lava flows and the scoria cones. The track to the caves runs through this moat for some distance. It was probably formed by the slight subsidence of the scoria cones into the vent as lava withdrew back down the volcano's throat at the end of Rangitoto's eruptions.
6. Summit. Superb views into the crater and out over the Waitematā Harbour, Hauraki Gulf and Motutapu Island. The small concrete building was the fire command post used to co-ordinate Auckland's coastal defence gun batteries during the Second World War.
7. Islington Bay Wharf and historic baches.
8. Causeway to Motutapu Island constructed on top of a shell spit during the Second World War to link defensive posts on both islands.
9. Old Harbour Board quarry and site of controlled mine base during the Second World War.
10. Wreck Bay. Reached via a rough side track (1 hour each way) off the road between Islington Bay and the summit. The low cliffs on either side of the bay and the huge boulders thrown up above high tide attest to the power of major storms that have carved into the basaltic lava flows on the exposed north side of Rangitoto since they were erupted just 600 years ago. The sheltered south side of Rangitoto has had almost no erosion.
11. McKenzie Bay, along with adjacent Whites Beach, is the only sandy beach on Rangitoto. It was the site of a short-lived (1892–96) salt refinery where imported rock salt was dissolved in sea water and then refined salt crystals were produced from the brine by evaporation in a huge pan heated by four furnaces. Nearby on an offshore islet is Rangitoto Beacon (built in 1887) – one of the first structures built on the island and used for sea navigation. It was initially lit by gas but switched to electricity in 1929.
12. Flax Point. Site of Auckland's largest black-backed gull colony.

Geology

Rangitoto is Auckland's largest volcano, having erupted more lava than all the others in the field combined. It is also the youngest of Auckland's volcanoes. Its fiery displays 620–600 years ago were the only eruptions of an Auckland volcano to be witnessed by humans. We know this because archaeologists have found the remains of a small Māori fishing village buried by a metre of Rangitoto ash on the shore of neighbouring Motutapu Island. There is no evidence that anyone in the settlement was killed by the eruption, but human and dog footprints preserved in layers of ash confirm that the local people were present during quiet periods between the eruptions.

The only one of Auckland's volcanoes known to have erupted in the sea, Rangitoto came up in the entrance to the Waitematā Harbour, in the middle of the main channel (the drowned course of the Waitematā River). Before it erupted, there was a deep-water channel all the way in to where downtown Auckland is now located. Because the eruption blocked the main channel, harbour authorities had to dig and deepen the present-day Rangitoto Channel through what is the drowned crest of a low ridge of Waitematā Sandstone.

Rangitoto began life with a major phase of wet explosive eruptions that powered fast-moving base surges of hot gas and ash across the sea where they plastered the west coast of nearby Motutapu Island. Large quantities of volcanic ash also erupted high into the air, with much of it blown northeast to mantle Motutapu Island with a metre or more of ash. It is inferred that coarser debris and ash built a tuff cone island in the centre of the harbour entrance, the trace of which is now buried beneath later lava flows and scoria deposits. Eroding beach deposits of Rangitoto ash have been found on the coasts of Motutapu Island (Islington Bay), Musick Pt, and Achilles Pt (St Heliers). Once the volcano's vent was

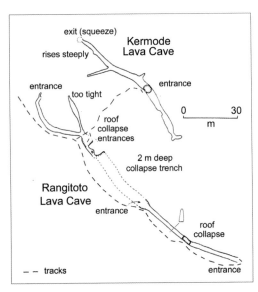

● Map of Rangitoto lava caves accessible via a short side track off the main walking track to the summit.

above water level and the sea was excluded from interacting with the rising molten magma, the eruptions switched to the dry style of fountaining eruptions and the voluminous outwelling of lava flows which built up the circular shield volcano of Rangitoto that we see today.

Geochemical studies by University of Auckland graduate student Andrew Needham in 2008 showed that two different batches of magma sourced from different depths within the mantle erupted to form Rangitoto. The first batch had an alkali basalt composition and came from about 90–100 km deep. Sediment cores taken from small swamps on Motutapu have 10–100-centimetre-thick alkali basaltic ash deposits separated by 30–50 cm of swamp silt from much thinner (less than 10 cm) volcanic ash with a subalkali basaltic composition, erupted from a second batch of magma. The northern scoria cone is of alkalic basaltic composition,

Part of the 100-metre-long Rangitoto lava cave. It is a narrow, straight, trench-shaped cave with several sections of collapsed roof that provide welcome light for those who explore it without a torch. The cave is reached along a short side track off the main walking track to the summit.

These narrow, branching lobes of pahoehoe lava flow were formed as the front of one of the younger flows entered the sea. They can be seen at low tide just over the sea wall from the information shelter at Rangitoto Wharf.

seemingly erupted from the first magma batch. The main and southern scoria cones and all of Rangitoto's lava flows were erupted during the second phase. The subalkalic chemistry of the second batch indicates that it was sourced from 65–75 km depth. It clearly came up the same conduit as the earlier batch and not too long after, as the route had not been blocked by cooling and solidifying magma. Solidified basalt had blocked the plumbing only just beneath the northern cone and thus this second batch of magma appears to have been rerouted a few hundred metres to the south to erupt at the surface.

It is hard to estimate the amount of time it would have taken for 30–50 cm of silt to accumulate in the Motutapu swamps between the two phases of eruption, as much material would have washed off the deforested hills in the first big rainstorms. Similarly, it is difficult to find plant material that may have colonised the new ash and was growing on the site when the second eruptions began and hence would have provided a reliable radiocarbon date of the time between eruptions. Further away, however, there are two thin Rangitoto ash layers (each less than 1 cm thick) in a core from the floor of Lake Pupuke. They too contain the two different compositions that Needham has recognised on Rangitoto and are separated by only a 1–2 cm thickness of lake sediment. Calculating the background rate of lake sediment accumulation indicates that there was a maximum of 20 years between the two phases of eruption of Rangitoto. Archaeologist Reg Nicol, who excavated the buried Māori village site on Motutapu Island, found evidence of gardening activity in the volcanic ash of the first eruptions prior to burial by ash from the second phase.

The second batch of magma probably erupted for several years, building a succession of scoria cones above the vent and extruding massive quantities of lava from all around the base of the growing cones. The youngest scoria cone forms the summit peak and has a substantial, 60-metre-deep crater. The remnants of earlier cones are partly buried around its sides.

Numerous overlapping lava flows gradually built up a gently sloping (about 10 degrees) circular shield volcano that surrounds the central scoria cones. Thus the characteristic profile of Rangitoto when viewed from Auckland City comprises the gently sloping shield on either side with three steeper bumps on the crest. The highest central bump is the youngest scoria cone and the lower two are partly buried earlier scoria cones, the northern produced during the first phase of eruption and the southern during the second phase.

Most of the lava flows were of the a'a variety, in which their molten lava slowly moved along within a thick carapace of broken-up clinkery crust. In many places the clinker crust is replaced by large slabs (up to 1.5 m across) of solidified basalt that probably formed on the surface of more fluid pahoehoe flows, but were later ripped apart and rafted along by the more slowly flowing and increasingly viscous lava that developed as the flow was cooling. These slab flows or blocky a'a flows are transitional in character between pahoehoe and a'a flows. Some lobes with ropey pahoehoe surfaces can be seen near the coast.

In cross-section most of Rangitoto is composed of gently sloping lava flows that extend down to below sea level. Only the very centre of

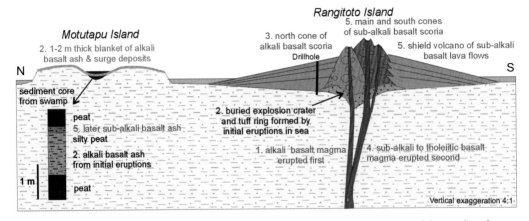

North–south cross-section through Rangitoto Volcano, summarising its sequence of eruption and the mantling of neighbouring Motutapu Island with wind-blown ash.

the island is formed of buried scoria cones, which probably grew up in concert with the growing lava-flow shield.

Drill core

In 2015, a core was drilled through the volcano, 1 km northwest of the summit and 120 m above sea level. It passed through at least 53 lava flows of subalkali basalt, each 0.5–7 m thick and totalling 128 m thick. There is no strong evidence for any long break in time between any of the flows. The flows overlie 7 m of scoriaceous tuff erupted during the early phase of activity and then 5 m of shell-bearing mud overlying much older Waitematā Sandstone. Shells within the marine mud have been radiocarbon dated, the results indicating that they were deposited in the harbour when the sea surface was close to its present level between 7500 and 650 years ago. Older shells (dated at 1600–4000 years old) occur in slivers of mud within the scoriaceous tuff pile and indicate that it was probably pushed along and mixed with the underlying mud by the emplacement of the overlying lava flows about 600 years ago.

A 1-metre-thick slab of basalt cored within 6000-year-old mud was initially interpreted as being a lava flow and taken to indicate a much earlier age for the start of eruptions at Rangitoto. More recent studies indicate that this basalt is the same age (600 years) and composition as the lower part of Rangitoto's main lava-flow sequence, and that it was likely a tongue of lava flow that was pushed down into the soft sea-floor sediment as the flows advanced into the sea while the island volcano was growing. Examples of similar shell-bearing mud that has been intruded by lava flows and pushed up out of the sea by them can be seen in the intertidal zone around the southwest coast of Rangitoto.

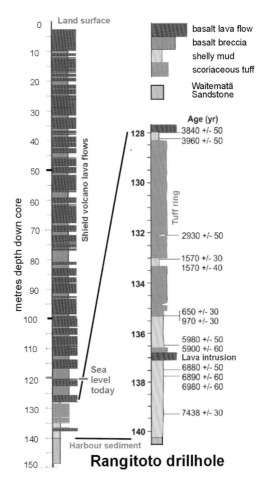

● Log of the drill core put down through Rangitoto Volcano in 2015. Column produced by MSc student Tamzin Linnell in 2016.

Groundwater

Rangitoto has no streams as all the rain that falls on the island either evaporates or soaks down into the highly fractured lava flows to become part of the groundwater. The fresh groundwater in Rangitoto's rocks has a slightly domed surface rising above sea level and is higher towards the centre of the island. It floats on top of a layer of denser salt water below sea level. Freshwater springs bubble out of fractures in the lava-flow rocks in many places around Rangitoto's coast, especially after periods of wet weather.

Vegetation

Botanically, Rangitoto is unique in New Zealand, with trees growing directly out of basaltic lava flows. The inhospitable basalt provides the substrate for the largest pōhutukawa (*Metrosideros excelsa*) forest in the world. The amount of vegetation cover is related to substrate. The crumbly scoria cones have already developed a thin soil layer and near-continuous forest cover. In contrast, parts of the lava flows are still bare, supporting only hardy lichens, while other parts are well forested by pōhutukawa, creating a mosaic of forest and open areas. The pōhutukawa usually colonised hollows on the surface of the lava flows and as the trees have grown they have formed 'vegetation islands', gradually advancing outwards over the inhospitable a'a lava as leaf litter accumulates under them. This litter has been colonised by mosses, ferns and other hardy trees and shrubs, including a number of plants that are more commonly found as epiphytes growing in the branches of trees, such as Kirk's tree daisy (*Brachyglottis kirkii*), northern rātā (*Metrosideros robusta*), puka (*Griselinia lucida*) and wharawhara (*Astelia banksii*). The sheltered southern coastline, and especially the muddy inlets between projecting lava flows, has been colonised by mangroves (*Avicennia marina*), many of which grow over solid basalt surfaces with their roots intruding down cracks.

Pre-European Māori history

Rangitoto can mean 'blood red sky' in Māori but there is no confirming tradition for this translation and, while it is attractive to think it refers to the eruption, there is an alternative. The name is traditionally linked with a fight between Tamatekapua and Hoturoa, the respective commanders of the *Arawa* and *Tainui* canoes. Tamatekapua was injured, hence the

◑ Austrian geologist Ferdinand von Hochstetter's sketch of Rangitoto made as he sailed into Auckland on the Novara Expedition on 22 December 1858. Hochstetter was the first to undertake a detailed geological study of Auckland's volcanoes but did not set foot on Rangitoto itself. *Courtesy of Hochstetter Private Collection, Basel*

◐ Painting depicting the early wet explosive ash eruptions from Rangitoto about 620 years ago. The thick ash buried a small Māori fishing village on the shore of neighbouring Motutapu Island. *Painting by Chris Gaskin*

name Te Rangi i totongia-a-Tamatekapua – 'the day the blood of Tamatekapua was shed'. Rangitoto's long use as a burial place confirms its great spiritual and symbolic importance to local iwi. Because of its rugged lava-flow surface, Rangitoto was not widely used by pre-European Māori, who naturally preferred adjacent Motutapu Island with its rich volcanic-ash soils, numerous flat areas and sandy beaches.

European history

The island shores provided an early source of basalt for use in building construction in Auckland, such as in the mission house across the water in Mission Bay and St Pauls in Symonds St. Rangitoto was designated a public domain in 1890 under the control of the Devonport Borough Council. The first track to the summit was constructed in 1897. The island's management

over the next 50 years had an impact on Rangitoto's naturalness, yet created a remarkable assemblage of historic cottages, or baches. By 1937, when the last leases were granted, there were 140 such buildings on the island, mostly around Rangitoto Wharf and Islington Bay. On the death of the lessee, most were demolished in the 1970s and 1980s, but in 1990 this policy was stopped. The Rangitoto Island Historic Conservation Trust was established in 1997 to restore and promote the historic heritage values of these unusual bach communities. Only about 30 baches remain today. Bach 38, near Rangitoto Wharf, has been restored and turned into a museum that is open periodically.

During the 1920s and 1930s, prisoners from Mt Eden jail constructed the roads and coastal tracks on Rangitoto. They also made the small sea-water swimming pool near Rangitoto Wharf,

By 1937, 140 small baches like these three near Rangitoto Wharf had been built around the southern shores of Rangitoto. Since then many have been demolished, but those that remain are now being restored and their historic heritage values celebrated. *Photo by Alastair Jamieson, 2010*

A side track off the summit road from Islington Bay leads to Wreck Bay on Rangitoto's north coast. The remains of at least 13 obsolete ships were dumped here and set on fire between 1887 and 1947. The bow of the steamer *Ngapuhi* can be seen in the middle of the bay. In just 600 years since their eruption the lava flows on this exposed northern side of Rangitoto have eroded back, forming 5-metre-high coastal cliffs.

the tennis pavilion (later the hall) at Islington Bay and the coastal stone walls around the landings at Rangitoto and Islington Bay wharves.

Building of structures that would contribute towards the coastal defences of Auckland began on Rangitoto and Motutapu islands in 1937 before the outbreak of the Second World War. During 1939–43, and especially when invasion by Japan seemed inevitable, both islands were closed to the public and hundreds of men and women were stationed there. A gun battery facing the channel between Rangitoto and Motukorea was built on the Rangitoto coast near the entrance to Islington Bay. Nearby on the coast of Rangitoto, a large controlled mine base was completed

in 1943. It was located in the former Auckland Harbour Board quarry that had been excavating basalt from lava flows since 1898. The mine base consisted of many large sheds, a wharf and powerhouse. The base was to service and detonate mines that were strung off the end of Whangaparāoa Peninsula and in Great Barrier Island harbours. After the war, the mines were removed and the base used for naval stores until it was demolished in the 1980s. In 1937, a fire command post was built on the summit of Rangitoto because of the uninterrupted views over the entrance to the Waitematā Harbour. Its purpose was to direct gunfire at attacking warships from coastal defence batteries around the harbour entrance. This concrete building still sits on the summit. Further around on the Crater Rim Track are the remains of several other structures from the Second World War – an observation post, wireless room, radar station and engine generator shed.

Pest eradication

By 1990, the native flora and fauna of Rangitoto were severely impacted by the browsing and predation by a plethora of introduced mammalian pests that were shared with adjacent Motutapu Island. The most severe damage was blamed on a population of about 2000 brush-tailed rock-wallabies and tens of thousands of brushtail possums, both of which had been deliberately introduced from Australia – the wallaby in 1873 and possum in 1931 and 1946. A campaign of aerial poisoning, bait stations and tracker dogs over both islands resulted in their complete removal by 1996. The canopy and understorey vegetation have recovered and flourished since that time, but there were still seven introduced mammalian pests on the islands that were having devastating impacts on the birds and terrestrial animals. From 2007 to 2011, Rangitoto and Motutapu were subjected to one of the most complex pest-eradication programmes in the world resulting in the removal of mice, ship rats, Norway rats, hedgehogs, stoats, rabbits and feral cats. Today, Rangitoto and Motutapu are pest free and every effort is made to ensure they remain that way.

Bird life

With rapid forest regeneration the native bird life has flourished, with many species returning without assistance. Today the island is home to tūī, fantail, grey warbler, silvereye, kākāriki, bellbird and even the large, forest-dwelling parrot, the kākā. Other native birds that have been transferred in are saddleback and whitehead, with takahē and brown kiwi released on Motutapu. While probably just visitors, takahē have been reported in the Islington Bay area of Rangitoto.

The main seabirds nesting on Rangitoto are Auckland's largest colonies of black-backed gulls, found on the open lava flows at Flax Point and north of Whites Beach. Little blue penguins also nest in places around Rangitoto's coast.

⊙ The steep-sided summit crater and viewing platforms on the crest of Rangitoto's highest scoria cone. Note also the rectangular roof of the concrete fire command post – a relic from the Second World War. *Photo by Alastair Jamieson, 2009*

Motukorea/
Browns Island

◉ Aerial view of Motukorea Volcano from the north showing some of the lava flows forming coastal reefs to the west (right) and south (beyond). The extent of sea erosion of the semicircular tuff ring in the last 7500 years can be gauged by the offshore reefs (left), which mark the former shoreline. *Photo by Alastair Jamieson*

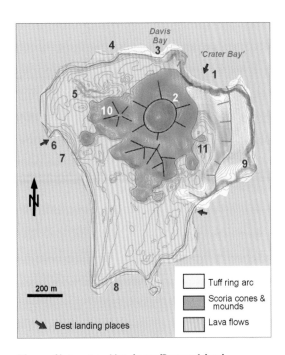

Places of interest on Motukorea/Browns Island:

1. 'Crater Bay'. The sea cliffs around this bay are made of bedded tuff capped by wind-blown scoria that fountained from the crater of the cone nearby. This bay was formed by sea erosion and is not a volcanic crater. It is the best sandy beach on the island.
2. Climb the scoria cone summit for superb views of Tāmaki Estuary and Strait and the crater.
3. Davis Bay. Site of a memorial cairn to Sir Ernest and Lady Davis, who gifted the island to Auckland City. Exotic trees and building foundations nearby mark the site of the Featherstones' house.
4. On the foreshore, a dark grey lava flow can be seen where it flowed out over the edge of the tuff ring.
5. A 2–3-metre-deep trench on the flat was formed by the collapse of the roof of a lava cave.
6. Old stone wharf.
7. At low tide you can see the remains of five old paddle steamers dumped on Motukorea's foreshore.
8. South end where an unusual mineral occurs as cement in the intertidal tuff rocks. It was first recognised here and named motukoreaite after the island.
9. East cliff top. In the long grass are the earthwork remains of a small pā site.
10. Small pā site on top of a secondary scoria mound.
11. Pre-European stonefield gardening area.

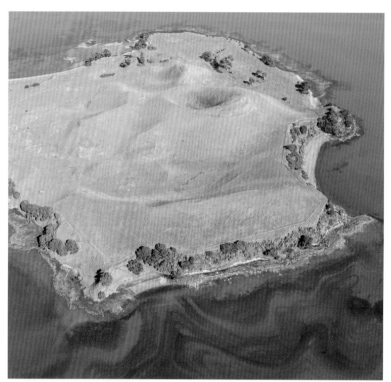

◔ The swirling pattern of eroded tuff beds on the sea floor off the east coast of Motukorea shows the former location of small stream valleys that were mantled by the earliest eruptions of volcanic ash about 25,000 years ago. Overlying tuff has been eroded by the sea in the last 7500 years. 'Crater Bay' is visible on the right. *Photo by Alastair Jamieson*

◔ The cliffs behind 'Crater Bay' on the north side of Motukorea are composed of bedded tuff overlain by loose dark red scoria. This bay is not a crater but was formed by marine erosion of the soft tuff ring in the last 7500 years.

Land status: Motukorea is administered by Auckland Council as a proposed regional park with full public access.

How to get there: There are no public ferry services to the island. The only way of getting there is by private boat, kayak, charter boat or water taxi. The best place to anchor or land at all tides is at the sandy beach ('Crater Bay') on the north side as there are many reefs around the rest of the island.

What to do: Most visitors climb up the track at the west end of Crater Bay and explore the island by ascending to the summit of the main scoria cone (30 minutes return) and by walking around the coastal fringe, which is easiest when the tide is not full.

Geology

Motukorea Volcano was formed about 25,000 years ago during the Last Ice Age, when the Tāmaki Estuary and Waitematā Harbour were forested river valleys. It began life with a series of wet explosive eruptions that cleared its throat of debris and created a 1-kilometre-wide shallow crater. The ejected debris ripped from the walls of the vent, together with large volumes of magma-sourced ash, accumulated around the crater to form a tuff ring. The eroding remnants of part of this tuff ring form the cliffs around the northern and eastern side of the island. At the time of eruption the wind was probably blowing from the southwest and a more substantial rim built up on the downwind side. Within the layers of tuff are numerous lumps of hard greywacke and Waitematā Sandstone. Both rock types underlie Motukorea at shallow depth and were pierced by the erupting throat of the volcano. Fragments of fossil shells also occur in the tuff layers, having been thrown out with the eruptions. These fossils came from a shell bed that partly filled a former Tāmaki Estuary channel that now underlies part of the volcano. Amongst the shells are specimens of the Sydney mud cockle (*Anadara trapezia*), which lives in estuaries around Australia today but disappeared from the New Zealand coast at the end of the Last Interglacial period about 120,000 years ago.

After the initial wet explosive eruptions, dry fire-fountaining built several scoria cones within

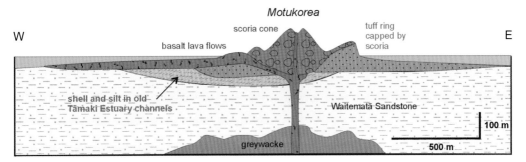

❍ East–west cross-section showing the distribution of rocks beneath Motukorea Volcano – the source of some of the lumps thrown out by the initial wet explosive eruptions and found in the bedded tuff.

the crater. Extrusion of lava flows from around the base of the cones appears to have rafted parts of their scoria ramparts away. These can now be seen as low mounds of scoria to the west and south of the main cone. The fluid basalt spread over the land forming an apron of flows extending 2 km to the west and south, temporarily ponding the Tāmaki Stream before it made itself a new course closer to St Heliers. These flows probably overtopped and removed the lower parts of the tuff ring around this sector of the volcano. The final phase of fountaining produced the high scoria cone with its deep summit crater in the centre of modern Motukorea. Some of the scoria was blown northeast and forms a 5-metre-thick cap on top of the tuff ring, seen in the upper part of the cliffs above Crater Bay.

As the ice caps melted after the end of the Last Ice Age, sea level rose to 1.5–2 m higher than at present by about 6500 years ago. The rising sea level submerged most of the lava flows. With no Rangitoto present to protect the island from storms, the sea rapidly eroded the soft tuff on the northern side, creating the high cliffs above Crater Bay. Much of this eroded material was swept around both sides of the island and together with shell built up the extensive flats on the south and west of the cone. The northern

◖ This 3-metre-deep trench across the northwestern flats of Motukorea was formed by roof collapse of a lava cave within a lava flow.

◖ The cream-brown clay-like mineral cementing these basalt pebbles together near high-tide level at the southern tip of the island is motukoreaite, which was first discovered here and named after the island in 1977. It has since been recognised in many places around the world.

⊙ Aerial view from the south of Motukorea's main scoria cone and central crater surrounded by numerous smaller scoria mounds formed by small secondary vents or by portions of scoria cone that were rafted away from the main cone by lava flows. Also visible is terracing and transverse ditches on the crest of the main cone and a secondary scoria mound (left). These are the remains of two pre-European pā. Stone mounds and walls on the floor of the eastern valley (right) are remains of pre-European gardens. *Photo by Alastair Jamieson*

extent of the eroded tuff ring is marked by a shallow reef that extends across the mouth of Crater Bay, 200 m from shore.

Human history

Motukorea, meaning 'island of the oystercatcher', was a prized occupation site in pre-European times for Ngāti Tamaterā. It was perfectly located for fishing and shellfish harvesting and the flats and gentler slopes were extensively cultivated, especially in the sheltered eastern valley between the central cone and the tuff ring. Here, stone heaps and low walls, relics from the gardening activities, can be seen on the valley floor when the grass is short. Stone-working areas have been found along the northwestern coast. Three defensive pā were constructed at various times

on higher parts of the island. The largest is on the crest of the main scoria cone and has defensive ditches and banks and numerous terraces. Earthworks from smaller pā can be seen on a small coastal headland on the eastern crest of the tuff ring and on the conical scoria mound west of the main cone.

Amongst the early European visitors to Motukorea was missionary the Rev. Samuel Marsden in 1820 who noted the Māori inhabitants were growing taro and kūmara in the rich volcanic soils. In 1827, however, when French explorer Jules Dumont d'Urville collected firewood from the island, he found it abandoned, possibly as a result of the musket wars.

In 1840, the 60-hectare island was purchased by Europeans William Brown and John Logan

Campbell from Ngāti Tamaterā, who at that time were located a considerable distance away in Coromandel. Brown and Campbell's canvas tents, raupō whare and pig run on the western flats were the beginnings of Pākehā commercial activities in Auckland. After just a few months, however, they moved to the newly established town of Auckland. In 1879, Motukorea was sold to the Featherstone family who farmed it and built a house on the northwestern flats. The house burnt down in 1915, some time after the island had been purchased by the Alison family in 1906. The site of this house is still marked by introduced trees, building foundations and a stone-lined well. The Alison family were principal shareholders of the Devonport Steam Ferry Company, and during their tenure of ownership Motukorea became popular for ferry excursions and picnics. The disintegrating hulks of five of their obsolete coal-powered paddle steamers can still be seen where they were dumped in the intertidal zone near the stone wharf on the western side of the island.

Two of the paddle steamer ferries – the *Tainui* (left) and *Alexandra* (right) – abandoned on the beach on the west side of Motukorea in 1908. Note the location of the farmhouse close to the stone wharf. *W. A. Price, Wikimedia Commons*

Skeleton of one of five obsolete paddle steamers that were abandoned on the west side of Motukorea by the Devonport Steam Ferry Company between 1908 and 1914. The remains of their hulks can still be seen at low tide near the old stone wharf.

◉ Motukorea/Browns
Island viewed from the
west showing the proposed
sewage treatment and
outfall scheme that was
abandoned and moved
to Māngere in 1954.
Weekly News, 1937

In 1931, the Auckland Metropolitan
Drainage Board began developing a plan to
pump Auckland City's sewage out to Motukorea
for rudimentary treatment before it would be
discharged in semi-raw state into the harbour.
Vociferous opposition to it progressively built
momentum amongst ratepayers in the eastern
suburbs and on Waiheke Island, with support
from the recreational boating fraternity. After
the end of the Second World War, planning
by the Drainage Board advanced rapidly and
the island was purchased in 1946. In 1944, the
scheme's opponents formed the Auckland and
Suburban Drainage League, led by a fiery local
businessman named Dove-Meyer Robinson
(later knighted after his long stints as mayor of
Auckland, 1959–65 and 1968–80). The bitter
battle lasted nearly 10 years and the scheme was
only overturned when 'Robbie' and his supporters
got themselves elected onto the Metropolitan
Drainage Board. By that stage contracts had
been awarded and work on the pipeline had
already begun at the foot of the cliffs near
Karaka Bay on the mainland. In its place, sewage
treatment was shifted to a new treatment plant
built at Māngere, with consequent detrimental
impacts on Puketūtū Island and Māngere
Lagoon volcanoes and the Manukau Harbour.

In 1955, Robbie's friend and brewery magnate
Sir Ernest Davis (mayor of Auckland, 1935–41)
purchased Motukorea from the Drainage
Board and gifted it to the city of Auckland for
a public reserve. The Auckland City Council
transferred the island's management to the
Hauraki Gulf Maritime Park Board in 1967,
and when the Maritime Park was abolished in
1990, the new Department of Conservation
inherited management on behalf of Auckland
City Council. In 2017, after a deliberately lit fire
scorched much of the island, its management
was taken back by the Auckland Council to
become a regional park.

Biota
The native forest cover of Motukorea would
have been removed in pre-European times.
Today it is mostly covered in dense kikuyu
grass, which until recently was grazed by cattle.
In 1991, the Department of Conservation
eradicated rabbits from the island and an
aerial poison drop in 2000 was successful in
removing the remaining pests – Norway rats
and mice. The island is home to New Zealand
dotterel and numerous oystercatchers.

Pupuke Moana/ Pupuke Volcano

Land status: Lake Pupuke is in public ownership and available for all to enjoy, with four lakeside reserves. The majority of the tuff ring that surrounds the lake is in privately owned and commercial properties, except for the roads, five of which run along the crest of the tuff ring – Shakespeare, Kitchener and Hurstmere roads, Killarney St and Lake Pupuke Drive.

What to do: Picnic by the lake in one of the parks; walk around the lake edge track in Sylvan Park and from there to Henderson Park; fish, sail, windsurf, row, canoe or learn to scuba dive in the lake. For more sustained exercise many people walk or run right around the lake on the roads that follow the crest of the tuff ring. Attend a show or have coffee at the old pumphouse and adjacent café in Killarney Park.

⊙ Aerial view from the north of the two coalescing explosion craters of Lake Pupuke (one large, one small on left) and surrounding tuff ring. *Photo by Alastair Jamieson, 2009*

Geology

Pupuke, at approximately 190,000 years old, is the oldest-known volcano in the Auckland Volcanic Field. The eruptive sequence of Pupuke Volcano differs from all the others in Auckland. Early activity appears to have been by dry eruptions from two vents building a low shield volcano of thin, overlapping basaltic lava flows. Lava of slightly different composition is inferred to have erupted from each of the two vents: one was near the centre of the present lake; the other in the northeastern corner where there is a semicircular embayment in the lake's otherwise circular outline. These dry eruptions possibly also produced small scoria cones directly above the vents, but any trace of these has gone, except near Northcote Rd where scoria deposits can be seen in road and quarry faces (see page 76). Fire-fountaining of frothy lava from an additional vent on the southwestern side of the basalt shield appears to have built up a small scoria mound that has been mostly removed by the former Smales Quarry.

The initial eruptions were followed by wet explosive eruptions that blasted out the large double crater that is now Lake Pupuke. The ash that was thrown out by these explosive pulses built up a tuff ring on top of the shield volcano. This tuff ring has steeper slopes on the inside than the gentler outer slopes.

The likely reason for this unusual sequence is that during the course of the eruptions water suddenly entered the active vents. The low lava-flow shield of Pupuke Volcano may have blocked the ancient Wairau Valley, which at the time drained south into Shoal Bay. This valley may have ponded and perhaps the fresh water overflowed into the Pupuke vents, reaching the rising molten magma and resulting in a series of violent explosive eruptions.

Pupuke Volcano has the only permanent freshwater crater lake (Lake Pupuke) now remaining in the Auckland Volcanic Field. The lake is roughly 1 km in diameter, 105 hectares in area and 57 m deep. The surface of the lake water is approximately 6 m ASL and its level is controlled by natural outlets around Thorne Bay (between Milford and Takapuna beaches) and Takapuna Reef. Here, large volumes of fresh water flow from the lake through underground cracks in the basaltic lava flows and discharge intertidally.

Like most of Auckland's volcanoes, Pupuke erupted during colder times of the ice ages when sea level was below the present and the Waitematā Valley and its tributaries were clothed in forest. Early lava flows from Pupuke invaded some of these nearby forests, and the fossilised shape of some of these trees and branches can be seen within the solidified basaltic lava flows in Takapuna Fossil Forest (see pages 77–81).

Pioneering Austrian geologist Ferdinand von Hochstetter visited Lake Pupuke in 1859 and was the first to describe its volcanic origins. His writings are the source of the oft-quoted myth that the water in Lake Pupuke comes through an underground aquifer from Rangitoto. These days we know that this is impossible and that all the lake water is purely derived from rain that falls on the lake and the surrounding inside of the tuff ring that slopes towards it.

Human history

The lake's name is a shortened form of Pupuke Moana, meaning 'the overflowing lake'. In early European times it was sometimes referred to as Lake Takapuna. The clear fresh water of the lake was used to provide the first reticulated water supply for Devonport Borough from 1894 when the first pumphouse was built. The old brick pumphouse still standing in Killarney Park was erected in 1906 and housed steam engines that pumped water to a reservoir on Mt Victoria. Additional pumps were later installed on the southwest side of the lake and used to supply water to Northcote Borough (1908–45) and

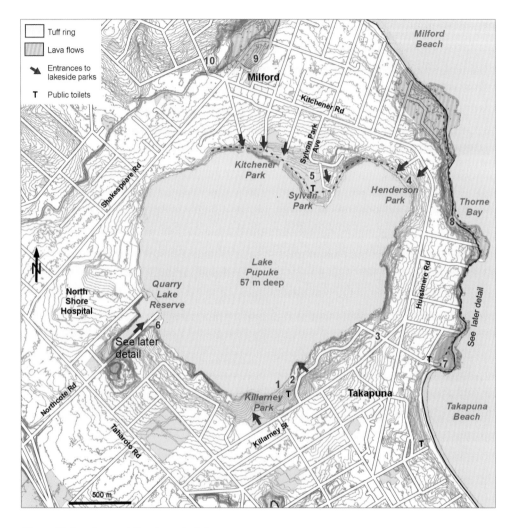

Legend:
- Tuff ring
- Lava flows
- Entrances to lakeside parks
- T Public toilets

Milford Beach

Milford

Kitchener Rd

Sylvan Park Ave

Shakespeare Rd

Kitchener Park

Sylvan Park

5
T

4
Henderson Park

Thorne Bay

8

North Shore Hospital

Quarry Lake Reserve

Lake Pupuke
57 m deep

Hurstmere Rd

See later detail

6

See later detail

3

Northcote Rd

1 2
Killarney Park
T

Takapuna

T 7

Takapuna Beach

Taharoto Rd

Killarney St

T

500 m

Places of interest around Pupuke crater lake:

1. Killarney Park. Car park, café, historic brick pumphouse (built 1906), The PumpHouse Theatre, jetty, ducks, picnic area.
2. Cliff exposure of tuff beside driveway down from Manurere Ave.
3. Crest of tuff ring at entrance to St Peter's Church. The Promenade has a steep slope on the inside of Pupuke crater and gentle outer slopes of the tuff ring run down to the sea.
4. Henderson Park. Car park, picnic area, jetty, start of grassy walk to Sylvan Park.
5. Sylvan and Kitchener parks. Car park, picnic area, boat ramp, lakeside walks in both directions. Pupuke Boating Club base.
6. Quarry Lake Reserve. Car park, North Shore Canoe Club and North Shore Rowing Club buildings and jetties, canoe polo pond. See Northcote Rd volcanic sequence on pages 75–76.

7. Takapuna Reef car park, end of The Promenade. Start of coastal walkway to Milford Beach, Pupuke lava flows, boat ramp, Takapuna Beach for swimming. See Takapuna Fossil Forest on pages 77–81.
8. Thorne Bay. Walking access only, swimming, freshwater springs outflow through rocks from Lake Pupuke. See Takapuna Fossil Forest.
9. Basaltic lava flows exposed in excavations in car park beneath Milford Shopping Mall.
10. Wairau Stream has a small waterfall flowing over lava flows from Pupuke just downstream from East Coast Rd bridge, Milford.

Birkenhead Borough (1913–34). From 1910 until the 1940s, Takapuna Borough also was supplied with water from Lake Pupuke. For most of the early 20th century, demand exceeded supply from the lake and water level dropped, at times reducing the size of the lake by 25 per cent and the lake level by up to 7 m lower than present and 3 m below high-tide level. As water quality in Lake Pupuke progressively declined, its use by local bodies was phased out by the 1950s.

In recent decades Lake Pupuke has become a popular venue for many recreational and competitive water sports – canoeing, kayaking, waka ama, dragon boating, rowing, yachting, windsurfing and scuba diving, even canoe polo in Quarry Lake, Northcote Rd.

Lake biota
Freshwater fish living in the lake include perch, tench, rudd, carp, catfish, goldfish and short- and long-finned eels. All of these, except the eels, have been introduced to New Zealand. Trout have been introduced to stock Lake Pupuke since the 1880s but as there are no streams entering the lake they are unable to breed here. In recent decades rainbow and brown trout have been reared at Fish & Game's Ngongotahā hatchery in Rotorua. Also present are small (1-centimetre-diameter) freshwater jellyfish that are native to China and were probably introduced from someone's freshwater fish tank.

Vast quantities of phytoplankton (freshwater algae) have lived in the water ever since the lake was formed. Much of the sediment on the floor of Lake Pupuke is composed of the tiny silica skeletons of diatom algal plankton. When the surface waters of the lake warm up in spring and summer the water may develop a pronounced brown discolouration due to blooms of tiny dinoflagellate algae.

In recent decades the edge of Lake Pupuke has been invaded by the introduced freshwater eel grass (*Vallisneria australis*). It lives attached by roots in the lake floor sediment at depths of 1–7 m and forms dense swards with 1–5-centimetre-wide strap-like leaves up to 5 m long. At times when the lake water level is lower, swans pull up the eel grass to eat the young shoots and discard the rest, which floats around rotting and becomes a boating hazard. In some summers the council has removed tonnes of rotting weed from the lake because of complaints about its smell and interference with boating activities.

↑ Lake Pupuke is a popular venue for budding young yachties and other water sports.

73

◑ Lake Pupuke pumphouse was built in 1906 to pump lake water to a reservoir on Mt Victoria for supply to Devonport residents. *Photographer unknown, 1927, Auckland Museum*
◑ Lake Pupuke from the west in 1930 with Shakespeare Rd in the foreground. Note the level of the lake water is well below the present (especially in the northeast bay at top left) as a result of three pumps taking water for the surrounding boroughs. *Whites Aviation Collection, Alexander Turnbull Library*

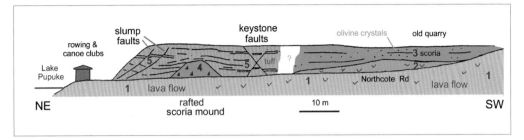

Points of interest in the Northcote Rd cutting:

1. Basaltic lava flows that formed part of the original Pupuke volcanic shield.
2. Remains of small hornito with basalt spatter buried within lava-flow sequence.
3. Layer of fine scoria, containing sparse crystals of green olivine, overlying lava flows.
4. Steep-sided, 5-metre-high mound of weathered scoria possibly rafted here on top of lava flow and now buried by layers of mantling tuff.
5. Bedded tuff with small arcuate slump faults and keystone faults resulting from compaction and slip in towards the crater (Lake Pupuke).

Northcote Road volcanic sequence

Where it is: In the road cutting on the south (right) side of the Northcote Rd extension that leads down to the North Shore Rowing and Canoe clubs on the western shore of Lake Pupuke.

The exposed rocks illustrate the sequence of eruptions that formed Pupuke Volcano. The base of the sequence is underlain by a dark grey basaltic lava flow, some of which was quarried away and now forms the drowned quarry on the west (other) side of the road that is used for canoe polo. The deep hole in Smales Quarry up the road on the east side was also formed by quarrying into this lava flow, which is part of Pupuke's early shield cone. Several thinner lava flows, separated by irregular horizons of broken basalt (breccia), form the uphill half of the road cutting and are also part of the shield. In one place a loose pile of twisted spatter surrounded by lava flow is probably the remains of a small hornito formed by eruptions of molten lava through a hole in the upper crust of the flow.

On top of these thinner flows, often hidden in the scrub above, is a layer of fine crumbly scoria that contains scattered light green crystals (up to 1 cm across) of olivine, a mineral that had crystallised out of the magma before it erupted and was ejected along with the semi-molten scoria in a dry fire-fountaining episode.

Overlying the lava flow and comprising most of the road cutting is bedded tuff erupted by the subsequent wet explosive eruptions. In one place there is a near-vertical-sided, 5-metre-high mound of scoria, which was probably part of a small scoria cone that was rafted along on the underlying lava flow. This hump has been draped and buried by the later wet tuff. Near the lake, small faults record a tendency for the tuff ring to slip down in towards the explosion crater. Further away they form criss-cross compaction patterns, or keystone faults. Displacements across the faults are generally only a few centimetres. Within the tuff there are a number of cobble-sized lumps of sandstone that were ripped from the wall of the volcano's throat during the eruptions. The tuff layers sag beneath several of these projectile blocks of sandstone, depressed by the impact of the rock landing in the soft, wet ash.

◐ Thin basaltic lava flows of the original Pupuke shield are the oldest parts of the volcanic sequence seen in the Northcote Rd cutting.
◐ A heap of twisted spatter buried within the Northcote Rd lava flows is thought to be the remains of a small hornito. Photo width 80 cm.
◐ Steep-sided knoll of scoria within Pupuke tuff ring exposed in the road cutting near the end of Northcote Rd.

Takapuna Fossil Forest and Takapuna–Milford Coastal Walk

The intertidal reef beyond the seaward edge of the car park at the north end of Takapuna Beach (end of The Promenade) has the best example in New Zealand of a forest killed and fossilised by passing lava flows. The forest of numerous small trees and a few larger kauri trees was growing here around 190,000 years ago at a time when sea level was considerably lower than at present. Takapuna Reef is composed of the remains of two thin lava flows from the initial eruptions of Pupuke Volcano that flowed into the forest and set it on fire. The first flow appears to have flowed right through the forest to a depth of 1–2 m, cooling and congealing around the tree stumps, forming cylindrical moulds composed of 0.3–0.6-metre-thick basalt. Most of the fluid lava between the moulds later drained away. As the tree moulds formed, the surface of the flow also crusted over in some places and some of this is preserved as arches between stumps that are close together. Most of the crust that formed on top of this first flow has disappeared, probably broken up by continued lava movement and carried off by the flow.

On parts of the reef there is evidence of a second, smaller flow of lava that followed the first. It has formed an additional, thinner solidified mould around many of the existing tree moulds. Less of the second flow drained away, as a solid sheet of basalt still fills the gaps between the tree moulds to a depth of 0.5–1 m in many places.

At high-tide level near the Takapuna Beach campground, there are depressions in the dark grey lava-flow surface that have been filled with the overlying orange-brown tuff, some of which has obviously been washed down into cracks and caverns within the flows themselves. This is the hardened remains of several metres of volcanic ash (tuff) that was subsequently erupted from Pupuke Volcano and buried the lava flows and fossil forest moulds. They remained buried and therefore protected from natural erosion and weathering until the last few thousand years when they have been uncovered by coastal erosion, which has removed the overlying softer tuff and left the lava flow and tree moulds as a hard intertidal reef.

⬆ Lava flowing through a forest created these cylindrical basalt moulds of the lower trunks of many small trees. Most of the tree moulds have hollow insides where the original wood slowly burned away.

**Six cross-sections through Takapuna Reef
illustrating the history of formation of the
fossil forest.**

1. Some 190,000 years ago, sea level was lower
   and a forest was growing where Takapuna Reef
   is now.
2. Pupuke Volcano began erupting and molten
   lava flowed through the forest, setting the
   trees on fire.
3. Lava cooled and solidified around the lower
   tree trunks, forming low basalt moulds. The top
   of the flows also solidified, forming a solid crust
   of basalt.
4. Remaining molten lava drained away, leaving
   behind vertical basalt tree moulds, some with
   arches between them.
5. Subsequent wet explosive eruptions from
   Lake Pupuke crater buried the moulds and
   remains of the lava flows.
6. In the last 7000 years, since sea level has
   risen to its present elevation, sea erosion
   has exhumed the tree moulds and lava flow
   remains, creating Takapuna Reef.

Illustration by Hugh Grenfell

⊙ Vertical view of part of Takapuna Fossil Forest showing
the cylindrical moulds of many tree stumps preserved in
basaltic lava. Over 200 tree moulds form most of Takapuna
Reef beside the boat ramp car park. *Google Earth, 2017*

A popular walking path follows the coast from Takapuna Beach north to Milford Beach. Just past the end of Brett Ave, there is a 1.5-metre-diameter metal cartwheel grill over the top of a large cylindrical hollow in the basalt rock. This prevents people falling into a 4-metre-deep mould of the lower trunk of a kauri tree that was growing here when Pupuke Volcano erupted. As with Takapuna Reef, the lava flowed around the tree trunk and solidified about it, preserving its shape before it had time to slowly burn away.

There was a shallow valley here and so the flow was thicker than at Takapuna Reef and it solidified in place. Because this lava flow was thicker, most of the smaller trees in this vicinity were completely engulfed and burnt without leaving a mould. The outline of a second, similar-sized kauri tree mould can be seen in the path, 3 m south of the grill-covered one. This one is full of ash and debris and the track runs over it. A third, upright kauri tree mould can be found in the lava flow off the end of Minnehaha Ave.

❺ This 1.2-metre-diameter trough is the mould of the fallen trunk of a kauri tree that has been preserved in the top of a lava flow at Takapuna Reef.
❻ These holes are moulds of four horizontal branches that were caught up in and rafted along by a passing lava flow from Pupuke Volcano. Before the wood was incinerated away, the lava solidified around them forming these tubular moulds. The mould second from right was still hot and plastic when a later lobe of lava flowed above and squashed it. The more irregular-shaped hollow (second from left) is a gas blister that formed within the cooling flow beneath its solidified upper crust. Visible beside the concrete sewer line between Brett and O'Neills avenues.

◔ This circular pool in a basaltic lava flow off the end of Minnehaha Ave is a mostly tuff-filled mould of the standing trunk of a kauri tree that was growing here 190,000 years ago when the lava from Pupuke Volcano swept around it.

◔ Overflow water from Lake Pupuke flows through joints in the basaltic lava flows all the way from the lake to emerge here at Thorne Bay on the coast.

Evidence of large gas blisters that formed
beneath the cooled surface crust of some of
the flows is also visible in this coastal section,
especially between the ends of Brett and O'Neills
avenues. As the flows were cooling and solidifying, small gas bubbles within the still molten
lava rose and coalesced with other bubbles to
form the blisters (1–3 m across and 0.5–1 m high)
that were trapped beneath the glassy surface
crust. On the ceilings of some of these blisters
are small lava stalactites (1–5 cm long) that were
formed by the heat of the trapped gas, which
partially re-melted the overlying basalt crust that
dribbled down before solidifying again.

◔ These small elongate globs of lava are stalagmites that
accumulated on the floor of a gas blister in a lava flow. They
have dropped off lava stalactites above while they were still
hot and plastic. Seen in the coastal rocks near the end of
Brett Ave, Takapuna. Photo width 20 cm.

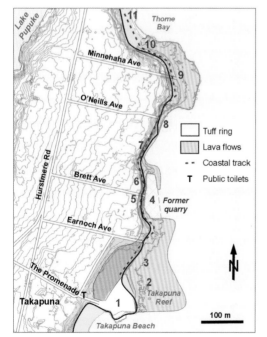

Places of interest on the Takapuna to Thorne Bay Coastal Walk:

1. Boat ramp and car park built over part of fossil forest
   in 1970s.
2. Takapuna Reef fossil tree stump moulds in lava flow;
   in places buried by brown tuff.
3. Mould of 1-metre-diameter felled kauri tree trunk.
4. Old intertidal quarry in lava flow.
5. End of Brett Ave. Moulds of horizontal branches in
   lava flow.
6. Upright kauri trunk mould under circular grill. Lava
   blisters with lava stalactites and stalagmites (10 m
   north of grill).
7. Alongside sewer line. Horizontal moulds of trunks
   and branches rafted along in lava flows. Also lava
   blisters and basalt cast of tree branch in mould.
   Flows covered by tuff higher in cliff.
8. 20 m south of end of O'Neills Ave. Segregation
   veinlets in basalt.
9. Opposite the path down from end of Minnehaha Ave.
   Shallow circular tide pool, 10 m out from path, is the
   mould of a vertical kauri tree trunk in lava flow. Mostly
   filled with brown tuff.
10. South end of Thorne Bay. Freshwater outflow from
    Lake Pupuke bubbles out through joints in intertidal
    basaltic lava flow. Water flows 250 m from the lake
    through underground cracks in the rock.
11. Coastal walking track to Milford Beach.

Te Kopua-o-
Matakamokamo/
Tank Farm/Tuff Crater

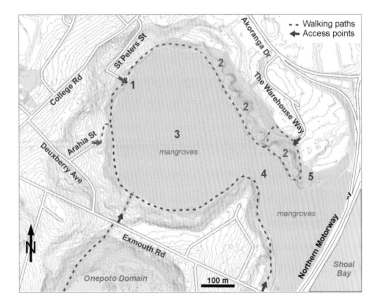

⊙ Tank Farm explosion crater and surrounding tuff ring in 2009 from above Shoal Bay in the southeast, showing its tidal channel passing under the Northern Motorway.
Photo by Alastair Jamieson

Places of interest at Te Kopua-o-Matakamokamo/Tank Farm/Tuff Crater:

1. Viewing platform and information.
2. Excavations for nine Second World War fuel tanks, now wetlands.
3. Mangrove and salt marsh fill the tidal lagoon in the crater.
4. Breach in tuff ring to the tide waters of Shoal Bay.
5. Area where tuff ring removed to help build motorway approaches to harbour bridge. Site of Forest & Bird's Millennium Forest.

● Photo of Tank Farm (left) and Onepoto (right) explosion craters and tuff rings, Northcote, during construction of the northern approaches to the harbour bridge in 1958. Note excavations removing parts of the tuff rings for fill. *Whites Aviation, University of Auckland*

Land status: The intertidal floor of the crater and a strip right around the coastal fringe is public reserve administered by Auckland Council. Most of the rest of the tuff ring is in private properties, except for the rim of the tuff ring that is part of three roads – Akoranga Drive, and College and Exmouth roads.

What to do: Take a leisurely walk around the semicircular shoreline track inside the crater (30 minutes each way) or return along the roads that follow the crest of the tuff ring. Spend an hour or so watching the rich bird life.

Geology

Tank Farm is a well-preserved, 800-metre-diameter explosion crater and high surrounding tuff ring that was formed by wet explosive eruptions soon after the neighbouring Onepoto and Lake Pupuke craters, about 180,000 years ago. These are thought to be the three oldest volcanoes in the Auckland Volcanic Field. Prior to the eruptions of Tank Farm and Pupuke volcanoes, the area was a shallow tributary valley of the Waitematā River that flowed down through what is now Shoal Bay. These two volcanoes dammed this valley, forming an extensive shallow lake that slowly filled with silt and peat, forming the flat area now occupied by the Wairau Valley industrial and commercial area. Overflow from the Wairau Valley lake was diverted naturally around the

northern edge of the Pupuke lava-flow shield. This new stream course spilt over the flows in a small waterfall, now beside Milford shopping centre, and escaped to the sea through the Wairau Estuary at the northern end of Milford Beach.

After its eruption the ~100-metre-deep crater filled with fresh water and became a lake with an overflow sill on its eastern side. This was not breached by the sea when sea level was ~6 m higher than present during the Last Interglacial period about 120,000 years ago, but the next time the sea rose to near-present level, about 7500 years ago, the salt water flowed over the sill and into the lake and erosion has subsequently widened and deepened the opening into Shoal Bay. As a result the crater filled with marine mud and became an intertidal lagoon. Unlike many of Auckland's breached explosion craters, Tank Farm still contains its original pre-human mangrove forest with a narrow fringe of salt marsh.

Human history

The Māori name for this volcano is Te Kopua-o-Matakamokamo, meaning 'the basin of Matakamokamo', derived from the traditional oral story of Te Riri-a-Mataaho or 'the wrath of Mataaho', the deity associated with volcanoes.

The name used by geologists is Tank Farm, a name that originated during the Second World War because of the fuel storage tanks that were dug into the north wall of the explosion crater to camouflage them from potential aerial attack by the Japanese. Early in 1942, the US Navy planned a major offensive against the advancing Japanese using Auckland as their base. This would require storage tanks for 75 million litres of fuel oil to supply the American fleet. Plans were approved to construct 50, 17-metre-diameter bolted steel tanks around the fringes of Tank Farm crater and Shoal Bay and link them by pipe to the Navy Base at Stanley Pt. Construction by the New Zealand Public Works Department, with materials supplied by America, began in mid-1942 but was called off in early 1943 as the Japanese advance had been reversed. By this time excavations and concrete bases for 25 tanks had been completed and six tanks erected and were in the process of being tested. Today, the circular depressions that remain at tide level inside Tank Farm crater are overgrown wetland habitats.

Many tank sites that were constructed around the Shoal Bay edge of the point north of the crater's mouth were destroyed when the northern approaches to the harbour bridge were built in the late 1950s. At the same time a portion of the tuff ring, where The Warehouse's head office now stands, was removed to provide fill for the motorway works.

◔ One of the nine circular holes dug in the northern edge of Tank Farm explosion crater during the Second World War in an attempt to hide navy fuel tanks from possible Japanese attack. Today most of these depressions are small overgrown wetlands.

In more recent years, the local North Shore Council relabelled this volcano Tuff Crater, a 19th-century descriptive term that was written beside it and a number of other explosion craters on Ferdinand von Hochstetter's 1864 map of Auckland's volcanoes. The North Shore branch of Forest & Bird adopted the crater for a major restoration project (Tuff Crater Restoration Project) that they began in 2000 with the planting of their Millennium Forest in front of The Warehouse building. Since then their team of volunteers has undertaken major animal and plant pest control, native plantings and track upgrades. They are currently proposing that the council link the ends of each track with a bridge and boardwalk across the tidal mouth to create a circular walk right around the inside of the crater, as already exists around Panmure Basin, Ōrākei Basin and Māngere Lagoon explosion craters.

Bird life

The reserve is home to many bird species, both native and introduced. The most common native birds seen are kingfisher, little and black shags, tūī, silvereye, kererū, Caspian and white-fronted terns, and banded and New Zealand dotterels.

Te Kopua-o-Matakerepo/
Onepoto Basin

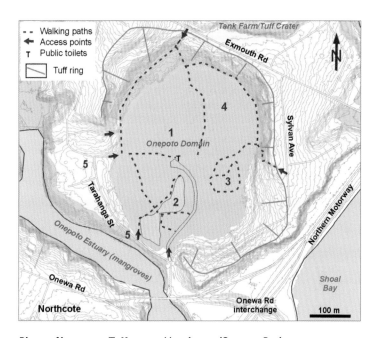

Places of interest at Te Kopua-o-Matakerepo/Onepoto Basin:

1. Flat floor of explosion crater filled in naturally with lake sediment and marine mud, capped with fill in 1975.
2. Two artificial ponds controlled by flood gates and used for sailing model boats.
3. Children's learn-to-cycle track and adventure playground.
4. Wetland boardwalk and bush loop track.
5. Area where tuff ring removed to help build motorway approaches to harbour bridge.

◑ Onepoto explosion crater and surrounding tuff ring, Northcote, from the west in 2009. Onepoto Stream (right) was diverted south around the new volcano to flow into the Shoal Bay tributary of the Waitematā Valley. *Photo by Alastair Jamieson*

87

Land status: The floor and most of the lower inner slopes of the crater are publicly accessible within Onepoto Domain, which is administered by Auckland Council. The rest of the tuff ring is in private property, except the crest, which is occupied by sections of three roads – Sylvan Ave, and Exmouth and Howard roads.

What to do: There are sports fields on part of the reclaimed land in the floor of the crater; two small lakes are popular for sailing model boats or watching eels. There is a children's adventure playground and an excellent sealed bike track for young children to learn to cycle on. Go for a barbecue or picnic. There is also a short boardwalk and bush walk loop around the northeastern part of the crater floor.

Geology

Onepoto is a 600-metre-diameter explosion crater and high surrounding tuff ring at Northcote that erupted about 185,000 years ago. When it erupted, sea level was lower than at present and mature kauri forest grew on this site. Hollow moulds of some of the trees that were killed and buried by volcanic ash were uncovered when the southwestern portion of the tuff ring (Tarahanga St area) was removed in the late 1950s to provide fill for the northern approaches to the harbour bridge.

At the time of eruption, Onepoto Stream was another arm of the Shoal Bay tributary of the Waitematā River. Initially the stream would have been dammed by the Onepoto tuff ring

⊙ View north along the line of three North Shore explosion craters in 1964. Northcote's Onepoto is in the foreground, then Tank Farm, with Lake Pupuke in the distance at Takapuna. *Whites Aviation, University of Auckland*

◑ Onepoto explosion crater and surrounding tuff ring was still largely unmodified by reclamation, roads or subdivisions when these two photos were taken in 1910. *W. A. Price, Alexander Turnbull Library*

but soon it eroded a new course around the southern side of the blockage. After the eruption the crater filled with fresh water to become a lake that overflowed over the lowest part of the tuff ring in the south into the recently diverted Onepoto Stream.

Geologists from the University of Auckland had two cores drilled near the centre of the crater floor in 2000–01. They passed through 36 m of marine mud, then 25 m of laminated silt that accumulated on the floor of a freshwater lake, before bottoming in the base of the crater at 61 m depth. This sequence of crater-fill sediment showed that the lake was not breached by the high sea level (6 m higher than today) during the Last Interglacial about 120,000 years ago. With more erosion of the overflow sill, the sea was able to flow up the Onepoto Stream valley and enter the lake as sea level slowly rose after the end of the Last Ice Age. After the breaching 8100 years ago, the freshwater lake became a saltwater lagoon for several thousand years as it rapidly filled with marine mud picked up from the Shoal Bay tidal mudflats and carried in suspension into the lake with every incoming tide. Once the mud filled up to mid-tide level it was colonised by mangroves and fringing salt marsh.

Human history

The most accepted Māori name for this volcano is Te Kopua-o-Matakerepo, meaning 'the basin of Matakerepo', as described in the tradition Te Riri-a-Mataaho. Another Māori name widely used today is Onepoto, meaning 'the short beach', named after a small beach (Halls Beach) near Northcote Pt.

Onepoto remained an intertidal basin with mangroves and salt marsh until 1975 when these were sacrificed and imported fill raised the level of the crater floor by about 1–1.5 m and turned much of it into playing fields for the rapidly growing population on the North Shore. It is currently the base for Auckland Australian Football League's North Shore Tigers Club. Two small artificial ponds were created near the natural outlet in the south with a flood gate to prevent sea water entering during spring high tides. One of these ponds has become the base for the New Zealand Radio Yacht Squadron, who races model sailing boats on it. In recent years there have been problems with flooding, siltation and pollution of the ponds that have been addressed by council but will probably recur from time to time.

Maungauika/
North Head

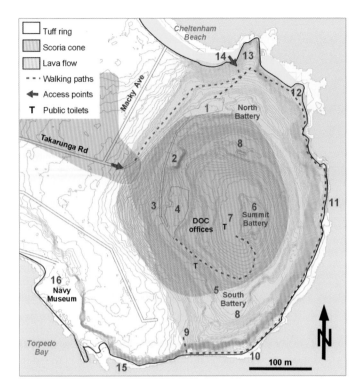

⊙ Maungauika/North
Head Volcano from the
southeast in 2018. The
lower parts are composed
of bedded tuff and the
upper third is scoria cone.
Photo by Alastair Jamieson

Places of interest at Maungauika/North Head:

1. Lower car park and entrance to tunnel network associated with North Battery (sometimes open to the public). Views to Rangitoto and Hauraki Gulf.
2. Old scoria quarry beside access road between lower and upper car parks.
3. Large spherical volcanic bomb beside access road to upper car park.
4. Upper car park.
5. Extensive tunnel network excavated in tuff and connected to South Battery and 13-ton disappearing gun. Saluting Battery of four 18-pound field guns overlooks harbour entrance.
6. Summit Battery. Panoramic views.
7. Army barracks and stone kitchen block (both built in 1885).
8. Second World War gun batteries.
9. Stairs and tunnel leading down to coastal walkway.
10. Wooden post and holes in coastal rock platform are remains of Second World War boom defence to stop enemy submarines entering the port.
11. Coastal walkway around base of North Head with excellent exposures of bedded tuff and searchlight and close defence gun emplacements.
12. Picnic area and sandy beach at foot of track leading down from lower car park.
13. Second World War gun position to defend Cheltenham Beach.
14. Stairway to Cheltenham Beach.
15. Old intertidal saltwater swimming pool, built in early 20th century.
16. Navy Museum, free admission.

⊙ Western side of North Head taken from Mt Victoria. *Hugh Boscawen, 1899, Auckland Museum*

Land status: North Head Historic Reserve is administered by the Tūpuna Maunga Authority. The Department of Conservation's North Head Office occupies some of the abandoned defence force buildings. The lower western and southern slopes are covered in privately owned houses.

What to do: Most visitors take in the views from the car parks or by walking around and up to the top of the volcano. Bring torches and explore the old fort's tunnels with the family or walk around the coastal track near tide level and back up through the access tunnel in the southeast corner of the reserve. Visit the Navy Museum (free admission) at the foot of North Head at the east end of Torpedo Bay (end of King Edward Parade).

Geology

Maungauika/North Head Volcano has been dated using the argon-argon method as erupting about 90,000 years ago, when sea level may have been a few tens of metres lower than at present. It began life with a series of pulsating wet explosive eruptions interspersed with drier eruptions of fine scoria. This ash and scoria built up a small tuff cone around a central explosion crater, which was completely filled and buried by the later scoria cone. The scoria cone vent was located in the northeast sector of the explosion crater and buried much of this side of the tuff ring. The southwest sector of the tuff ring is not covered in scoria and its flattened crest forms the terrace under Jubilee Ave.

Scoria and ragged scoriaceous lumps form only the top third of the cone. As the final scoria and lava bombs were erupted from the vent, a small stream of lava flowed westwards from its base and now lies hidden beneath the houses in Takarunga Rd. A small crater used to be present on the western side of the cone's summit. Ferdinand von Hochstetter visited North Head in 1859 and described its slopes as covered in 'volcanic bombs found on its surface, regularly pear-shaped or lemon-shaped bombs with their apices spirally turned'.

Marine erosion has eaten back into the lower parts of the volcano exposing the tuff beds in the fringing sea cliffs. The outward-sloping tuff beds form the shore platform around the southern, eastern and most of the northern sides

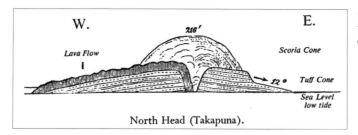

of the headland. Near the summit of the cone,
underground tunnels and bunkers in the layers
of loose and partly welded scoria were dug as
trenches and lined with concrete and a concrete
roof constructed, which was then buried.

For a short while, about 7000 years ago, following sea level rise to its present height after the
Last Ice Age, North Head was probably an island.
However, within a few thousand years, a sand and
shell spit was thrown up across the shallow gap
between Cheltenham Beach and Torpedo Bay,
linking North Head to the rest of the North Shore.

➊ Layers of volcanic ash (tuff) in the coastal cliffs of North
Head were deposited by base surges and air fall. The light-
coloured rock fragments were ripped from the wall of
the volcano's throat and thrown high into the air by the
explosive blasts.
➋ A 1-metre-diameter spherical breadcrust bomb beside
the driveway leading to the upper car park, North Head.

Pre-European Māori history

North Head was traditionally known as Maunga-
a-Uika – 'the mountain of Uika' – after an
ancestor who occupied it about 800 years ago.
He was a young relative of the famous Māori
voyager Toi-te-huatahi. The name Takapuna has
also been used for the cone, although it traditionally referred specifically to a spring flowing
from its base. The hill was weakly terraced by
pre-European Māori, although evidence of clear
defences from this period has not been found.

When the Navy Museum was being developed
on the low terrace at the southern foot of the
volcano, archaeologists uncovered one of the
oldest Māori occupation sites in Auckland. This
former coastal campsite contained cooking ovens,
charcoal, moa bones and a shell midden all buried
by an earth flow and later reclamation. Today this
site lies beneath new tarseal of the car park.

◔ The two oldest buildings on North Head are the army barracks (left) and stone kitchen block, both built in 1885 and later used to house prisoners constructing the tunnel networks around the north and south batteries.
◔ All the tunnels servicing the north and south batteries on North Head were excavated into bedded tuff and some are not lined with concrete, allowing the tuff layers to be examined.

◔ Remains of a coastal searchlight emplacement near sea level at the foot of North Head, 2010. It was installed for use in locating enemy ships attempting to enter the harbour at night during the Second World War.

Fort Cautley

The cone's strategic location at the entrance to the Waitematā Harbour meant that it was an important coastal defence site from the earliest colonial period. The size, variety and time span of its fortifications make it the most important coastal defence heritage site in New Zealand.

In the 1840s, North Head became the first pilot station for ships entering the Waitematā Harbour. In 1878, it was set aside as a reserve for defence purposes, if needed. Soon afterwards, fears of Russian attack on British Pacific territories led to the construction of a muzzle-loader fort (1885), and three 8-inch disappearing gun batteries (1886). The northern battery covered Rangitoto Channel, the south battery covered the main channel into the inner Waitematā Harbour and the summit battery covered all the area in between. Each battery had a giant, 13-ton disappearing gun and one is still on display in the South Battery.

Fort development continued through the late 19th century with installation of observation turrets, a generator and searchlights. Between 1888 and 1914, the summit army barracks building (1885) housed prisoners who excavated many of the tunnels that link the batteries and their service facilities. Further batteries were embedded in North Head during the first decade of the 20th century. For a time, North Head was known as Fort Cautley, named in 1885 after Major Cautley of the Royal Engineers, but this name has now lapsed and has been transferred to the naval establishment in nearby Vauxhall Rd.

Land at the southern foot of North Head in Torpedo Bay was reclaimed for a naval yard and mine storage in the late 19th century. Between 1892 and 1908, mines were strung across the harbour entrance between North Head and Bastion Pt and were to be electrically detonated from onshore if an enemy ship attempted to enter the harbour.

A second phase of tunnel construction and defence development occurred on North Head leading up to and during the Second World War, 1937–45. Two 4-inch gun batteries and a fire control post were constructed and close defence guns were installed around the base of North Head. Booms with anti-submarine nets were strung from North Head to Bastion Pt at this time.

North Head fort was decommissioned in 1958 and most became a reserve within the Hauraki Gulf Maritime Park in 1972. The navy finally withdrew completely in the 1990s and some of its remaining buildings were taken over for use by the Department of Conservation. Many other buildings have been removed.

◑ Aerial view from the south across North Head showing the defence buildings and gun emplacements in 1958 at the time the fort was decommissioned and most land opened up to the public as a reserve. *Whites Aviation, University of Auckland*

Takarunga/Mt Victoria

Land status: Most of Mt Victoria scoria cone is in a recreation reserve administered by the Tūpuna Maunga Authority. Some of the lower slopes and most of the lava flows to the south are covered in private dwellings. The foreshore lava flows between Devonport Wharf and Torpedo Bay are easily accessible.

What to do: Walk to the top of Mt Victoria scoria cone for spectacular panoramic views in all directions – one of the best in Auckland. On a calm day, it is a wonderful place for a picnic. On a windy day, bring the kids to fly a kite. Explore the remains of European fortifications around the summit. Walk along the Devonport foreshore (King Edward Parade) across the toes of Mt Victoria's lava flows, seen in the dark grey intertidal rocks.

◉ View in 2009 from the north across Mt Victoria's scoria cone with its U-shaped crater breached to the south (far side). Note the pre-European terraces on the northeastern slopes. *Photo by Alastair Jamieson*

◉ View east from the southern slopes of Mt Victoria in 1879 with the rafted scoria mound of Duders Hill on the Devonport foreshore (right). The hill was quarried away in the early 20th century. *Auckland Museum*

◉ View from the south over Devonport with Mt Victoria behind, 2009. All the dwellings between the volcano and the coast are built over rocky lava flows that came out from the breached crater. The foreshore rocks are the toes of these flows. *Photo by Alastair Jamieson*

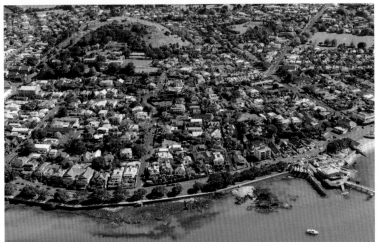

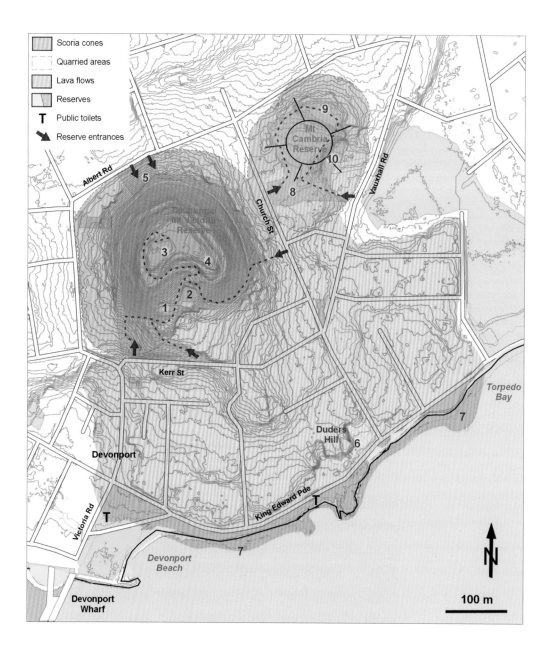

❂ Places of interest at Takarunga/Mt Victoria and Takararo/Mt Cambria:

1. Michael King Writers' Centre on Summit Rd in the former signalman's house, built in 1898.
2. Old tennis courts located in scoria cone's semicircular breached crater. They were constructed in the 1930s or earlier.
3. Mt Victoria summit. Panoramic views, flat top of historic water reservoir with 'mushroom' air vents, signal station tower (1964), 1899 disappearing gun and pit.
4. Bunker built in 1891 as a fire commander's post. Part of the North Head, Fort Takapuna and Mt Victoria defences of the Waitematā Harbour. Used for many years by the Devonport Folk Music Club for its regular meetings.
5. Historic Mt Victoria cemetery contains the graves of many early Devonport residents, 1856–1930s.
6. Former location of Duders Hill, named after Thomas Duder, signalman on Mt Victoria from 1843 to 1875.
7. Dark basaltic lava-flow toes exposed intertidally along the foreshore from near Devonport Wharf to Torpedo Bay.
8. Mt Cambria Reserve. Remnant hump of fused scoria, too hard to quarry. Square hole was the explosives locker when the quarry was active.
9. Highest remaining point on circular walk around quarried-out Mt Cambria scoria cone.
10. Devonport Museum, specialising in the history of Devonport; open to the public most afternoons, entry by donation. The museum and car park are backed by steep quarry faces of loose scoria, part of Mt Cambria scoria cone.

❂ Mt Victoria (centre), Mt Cambria (right) and Duders Hill (far left) viewed across Torpedo Bay, Devonport, in 1925. *James Richardson, Auckland Museum*
❂ The same view in 2010, except that Duders Hill and Mt Cambria are no longer visible as they have been quarried away.

99

Geology

Eighty-two-metre-high Takarunga/Mt Victoria dominates the Devonport landscape and provides visitors with spectacular views over the Waitematā Harbour. The cone was formed by fire-fountaining of frothy scoria from a central crater about 35,000 years ago. Towards the end of these eruptions lava flowed out from the southern base of the growing volcano, creating a fan of rocky basaltic flows between the mountain and the Devonport foreshore. The lava stream carried away some of the scoria cone on the south side, creating a breached crater, which the road to the summit (built in 1925) winds up through. The scoria that was rafted away came to a stop near the south end of Church St where it stood as a small mound known as Duders Hill, until it was quarried away in the early 20th century.

Human history

The Māori name for this volcano is Takarunga, meaning 'the hill standing above', with reference to its height above adjacent Takararo/

Mt Cambria. This was an important Māori pā for many generations and earthworks are still visible on some of its slopes. The European name of Mt Victoria comes from Victoria, queen of the United Kingdom 1837–1901.

French navigator Dumont d'Urville climbed Mt Victoria in 1827, and the strategic importance of the mountain was recognised when the first signal station for the port of Auckland was erected on its summit in 1842, giving it the early European name of Flagstaff Hill. Flags were used to signal to boats entering the harbour and inform city residents of incoming ships. Mt Victoria is still used as a signal station by the Ports of Auckland. The new signal station tower, constructed in 1954, is now fully automated with radar coverage of the harbour entrance from the Hauraki Gulf and monitored by Ports of Auckland across the harbour.

In the early 1880s, rumours spread that the Russians had their eyes on the South Pacific and before too long their navy would be attacking New Zealand. North Head and Mt Victoria, the two prominent scoria cones at the south end of

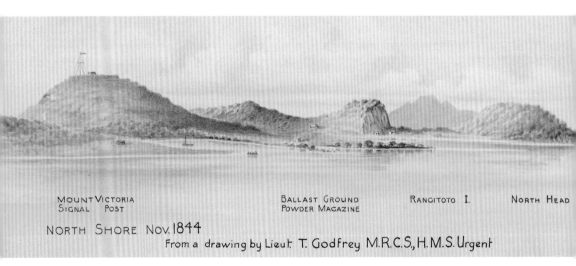

⊙ Mt Victoria, Mt Cambria, Duders Hill, Rangitoto and North Head, 1844. *Lt T. Godfrey of HMS Urgent, Auckland Art Gallery*

the North Shore peninsula, were perfectly placed natural high points for use as naval forts to defend the entrance to the youthful city of Auckland. So it was in 1885 that Mt Victoria was fortified with four 64-pound rifle muzzle-loader guns on a terrace on the north side and the summit became the observation point and control post for Auckland's coastal defences. In 1899, these guns were replaced by a 13-ton breach-loading disappearing gun that was hauled to the top and installed in the underground complex that still exists beside the summit car park. These guns were designed to fire a shell and then withdraw below an iron shield for reloading. The gun could fire shells 8 km (as far as Rangitoto) but was fired only once (for practice). The vibrations from the firing cracked windows in buildings in Devonport. Fort Victoria was not used for defensive purposes during the First World War but was utilised for a while to store ammunition. During the Second World War some of the bunkers were reopened and four anti-aircraft guns were mounted on the summit alongside an observation post.

In 1892, Devonport Borough Council constructed a water reservoir on the summit of Mt Victoria. This was used to store fresh water pumped there from Lake Pupuke and provide a gravity feed to the surrounding settlement. The buried reservoir and its 1911 extension still exist beneath the distinctive red-and-white 'mushroom' air vents, creating the artificial flat summit of the volcano. Lake Pupuke water has been replaced by water from dams in the Waitākere Ranges since the 1950s.

Grazing of stock on Mt Victoria occurred from the 1850s until about 1971 when the practice was discontinued. The road to the summit was completed and opened to public vehicles after the Second World War. In 2018, vehicle access to the summit was discontinued by the Tūpuna Maunga Authority except in special cases for the physically handicapped and elderly.

◑ The flat summit of Mt Victoria hides a buried water reservoir (1892, 1911) with mushroom vents above, 2010. The fully automated signal station tower beyond provides radar coverage of the entrance to the Waitematā Harbour.
◑ It is still possible to view the 13-ton disappearing gun that was installed on the summit of Mt Victoria in 1899.

Takararo/Mt Cambria

Land status: The quarried-out stump of Mt Cambria is now a reserve administered by Auckland Council. Remnant lower slopes of the scoria cone are covered in private dwellings.

What to do: Walk around the perimeter of the old quarry on a sealed path; bring the family for a ball game on the grass or a picnic. Devonport Museum is located in part of the former scoria quarry on the east side.

Geology

Mt Cambria was once a 40-metre-high scoria cone with a perfect crater and two high points on its rim. It stood on the site of present-day Mt Cambria Reserve between Church St and Vauxhall Rd in Devonport. More than a century of quarrying (prior to 1978), initially by private operators and later by Devonport Borough Council, has largely removed it and the quarry site has been landscaped as a pleasant neighbourhood reserve. Remnants of the lower slopes of the scoria cone underlie houses to the north of the reserve. Austrian geologist Ferdinand von Hochstetter reported a small lava flow sourced from its southeast flank but there is no visual evidence of this today. There is also no sign of any tuff ring or explosion crater that may or may not have been formed prior to the dry fountaining phase that produced the cone. Mt Cambria has not been precisely dated but argon-argon dating gives an eruption age of somewhere between 30,000 and 50,000 years.

◔ Mt Cambria Reserve (bottom left) from the north, 2018, with Mt Victoria beyond (to right). Note how rapidly the planted vegetation has grown since 1990. *Photo by Alastair Jamieson*

◉ Mt Victoria and Mt Cambria scoria cones sketched from North Head in 1872. Note the two high points on the circular rim of Mt Cambria's fire-fountaining crater. *James D. Richardson, Sir George Grey Special Collections, Auckland Libraries*

Human history

The Māori name for this volcano is Takararo, meaning 'the hill below', which refers to Mt Cambria being the smaller cone below adjacent Takarunga/Mt Victoria. Takararo was the site of a pre-European Māori pā but all of its earthworks have been quarried away along with the cone. Hochstetter visited the volcano in 1859 and named it Heaphy Hill after his helpful local assistant and good friend at the time, Charles Heaphy; however, this name has not lasted. The land was purchased by a syndicate of Welsh owners who named it Mt Cambria in 1874 after their native homeland and a gold mine they had owned in Thames, Coromandel, with the same name.

◔ Mt Cambria from Mt Victoria. *Photographer unknown, 1907, Wikimedia Commons*
◔ Mt Cambria Reserve from the slopes of Mt Victoria, 1994.

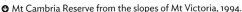

Quarrying began in the 1870s and continued through to 1978. In the latter decades of quarrying the Devonport Borough Council works' depot was also located here. After closure, the quarry and depot site were rehabilitated as a grassy reserve that opened to the public in the early 1990s. The Devonport Museum, run by the Devonport Historical and Museum Society, is housed in part of an early Presbyterian church moved in 1978 into part of the former quarry from where it was located on the corner of nearby Church and Cracroft streets.

Places of interest at Takararo/Mt Cambria:
See pages 98–99.

Volcanoes of central Auckland

Fifteen volcanoes are present in the central isthmus of the Auckland Volcanic Field. Many have large iconic scoria cones that act as centrepieces for, and sometimes gave a name to, Auckland's inner suburbs (Mt Eden, One Tree Hill, Mt Albert, Mt Roskill, Three Kings, Mt Hobson and Mt St John). The scoria cones of four of the volcanoes have been quarried away (Albert Park, Little Rangitoto, Mt Smart and Te Pou Hawaiki) and most of the cones of Three Kings have also gone. Five of these central volcanoes are recognised by their large explosion craters – Ōrākei Basin and Te Hopua craters have been breached by the sea, whereas Three Kings, Auckland Domain and Grafton craters are located on ridges. Many of these volcanoes erupted large volumes of lava from around the base of their cones. These extensive rocky lava-flow fields, with their rich, warm volcanic soils, were prized for gardening by pre-European Māori. In early European times they were converted to pasture and the loose basalt used to construct drystone walls between paddocks. Some of these walls survive as property boundaries within suburbia. Today, all these volcanoes or their quarried sites, except Te Pou Hawaiki, are partly or fully protected reserves.

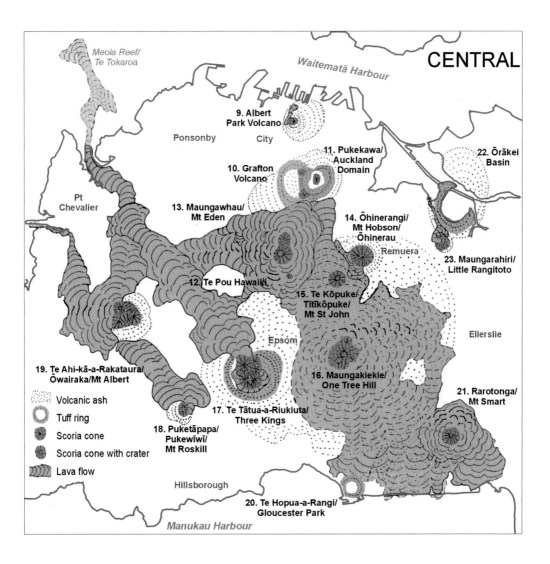

The scoria cones of Te Kōpuke/Mt St John (foreground) and Ōhinerangi/Mt Hobson (middle) are two of the much-loved volcano reserves of Auckland's inner suburbs. *Photo by Alastair Jamieson, 2018*

Meola Reef/
Te Tokaroa

Waitematā Harbour

CENTRAL

9. Albert
Park Volcano

Ponsonby

City

11. Pukekawa/
Auckland
Domain

22. Ōrākei
Basin

10. Grafton
Volcano

Pt
Chevalier

13. Maungawhau/
Mt Eden

14. Ōhinerangi/
Mt Hobson/
Ōhinerau

Remuera

23. Maungarahiri/
Little Rangitoto

12. Te Pou Hawaiki

15. Te Kōpuke/
Tītīkōpuke/
Mt St John

Ellerslie

Epsom

19. Te Ahi-kā-a-Rakataura/
Ōwairaka/Mt Albert

16. Maungakiekie/
One Tree Hill

21. Rarotonga/
Mt Smart

Volcanic ash

Tuff ring

Scoria cone

17. Te Tātua-a-Riukiuta/
Three Kings

Scoria cone with crater

18. Puketāpapa/
Pukewīwī/
Mt Roskill

Lava flow

Hillsborough

20. Te Hopua-a-Rangi/
Gloucester Park

Manukau Harbour

105

Albert Park Volcano

↑ There is no surface trace of the small Albert Park Volcano near the heart of Auckland City. It was centred between the old Magistrates' Court (Kitchener St) and the Victoria St car park building in the centre of the photo, at the foot of Albert Park sandstone ridge. *Photo by Alastair Jamieson, 2009*

Land status: The sites of the scoria cone and lava flow are covered by commercial buildings and roads.

Where it was located: Between Albert Park and Shortland St in the vicinity of the old Magistrates' Court with the flow extending down to Queen St.

Albert Park Volcano has a misleading name, as it was not located in Albert Park. Today, there are no clear landform remains of this small volcano, with most of what is known about it coming from the brief description by Ferdinand von Hochstetter during his 1859 visit and from subsurface geological evidence encountered during excavations for downtown buildings.

This was a small volcano that erupted about 145,000 years ago in a valley between the sandstone ridges near the heart of present-day Auckland's central business district. It began with wet explosive eruptions that deposited a veneer of volcanic ash up to 8 m thick over much of the nearby surrounding land, including the pre-volcanic sandstone ridge that underlies Albert Park and Auckland Art Gallery on its southwestern corner.

Forest surrounding the vent was killed and buried by this ash, as seen in an excavation near the art gallery in the 19th century. Later, eruptions switched to dry fountaining style, which built up a small scoria mound in the vicinity of the vent. A little scoria landed on the northern slopes of Albert Park and can sometimes be seen eroding out of the bank adjacent to one of the sets of steps leading up from Bowen Ave. A small lava flow oozed from the western base of the cone and flowed for several hundred metres down the Queen St valley floor. This flow dammed the stream, forming a swamp that now underlies Queen St between Victoria and Wellesley Sts and causes foundation problems for buildings. Hard basalt from this flow is sometimes encountered in building excavations north of Victoria St. The scoria was quarried away to help build the roads of the central city area in the first few decades after the establishment of Auckland in the middle of the 19th century.

Albert Park, after which this volcano has been named, was itself named after Prince Albert, Queen Victoria's consort. Prior to the arrival of

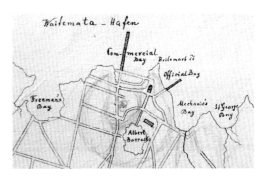

◐ This sketch of central Auckland by Ferdinand von Hochstetter in 1859 suggests that Albert Park Volcano (shaded in pencil) had a large eroded tuff ring spanning lower Queen St valley with a small scoria cone on the east side. *Courtesy of Hochstetter Private Collection, Basel*
◑ By 1869 there was no sign of any remains of the low Albert Park scoria cone that had stood in the very centre of this scene near the corner of Kitchener St and Courthouse Lane, Auckland City. *James D. Richardson, Sir George Grey Special Collections, Auckland Libraries*

European settlers, the park was home to the Māori village of Rangipuke. In the 1840s–60s it was the site of part of the 9-hectare Albert Barracks, which housed up to 900 imperial troops until their withdrawal from the colony in 1870. Part of the barracks' wall still stands near the University of Auckland library building. In the 1870s, Albert Park became a public reserve that was developed from 1882 onwards as a formal park and gardens by Auckland City Council using the competition-winning design of architect James Slater.

Grafton Volcano

⊕ Oblique aerial view of the site of Grafton Volcano between Outhwaite Park (small green reserve above railway line on right) and Auckland Hospital (white multi-storey buildings on left). View from the west, 2018. *Photo by Alastair Jamieson*

Land status: The site of Grafton Volcano is mostly covered in commercial properties, the University of Auckland medical school and Auckland Hospital. The only public reserve is small Outhwaite Park.

Where it was located: Much of the suburb of Grafton was built on the flat floor of the filled crater and its surrounding slopes margined by Auckland Hospital, Auckland Domain, Khyber Pass and upper Grafton Rd.

In Grafton there is little sign today of one of Auckland's older volcanoes, thought to have erupted just before the Domain Volcano about 100,000 years ago. This is because it was largely buried beneath volcanic ash from the slightly younger Auckland Domain explosion crater, which is inferred to have erupted through and blasted away the eastern arc of Grafton Volcano's tuff ring. Until recently the composition and extent of this volcano were little known and, because it is mostly hidden, our understanding is still somewhat speculative.

Numerous boreholes in the Auckland Hospital and adjacent medical school grounds, together with geophysical surveys by University of Auckland graduate student Sian France, have shown that the area between Outhwaite Park and Auckland Hospital is underlain by 30 m or more thickness of solid basalt buried by 3–15 m

of Domain tuff. This basalt appears to be a former lava lake within a 300-metre-diameter explosion crater with surrounding tuff ring. A thick mantle of tuff over upper Khyber Pass Rd and Symonds St probably came from the wet explosive eruptions from this centre.

A water borehole in the hospital grounds cored over 90 m depth of basalt, which suggests the presence of a former lava-filled vent (plug) in this area. Explosive fountaining probably threw up small scoria mounds from two vents – one under the hospital and another near present-day Outhwaite Park. The highest point of this latter scoria mound, mantled by Domain tuff, forms Outhwaite Park. This small reserve is named after the settler family of Thomas and Louisa Outhwaite, who purchased the property in the 1840s and lived there for 80 years. The park was donated to the citizens of Auckland.

Places of interest around Grafton Volcano:
See pages 110–11.

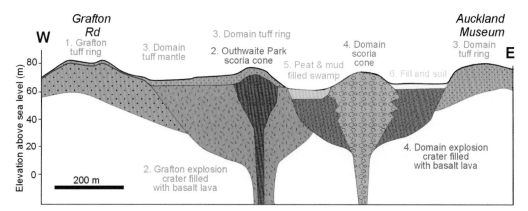

⊙ Schematic east–west cross-section through Grafton and Auckland Domain volcanoes, summarising the sequence of eruptions (1 to 5). The Domain explosion crater erupted through the east side of Grafton explosion crater and tuff ring and buried it beneath a thick mantle of volcanic ash.

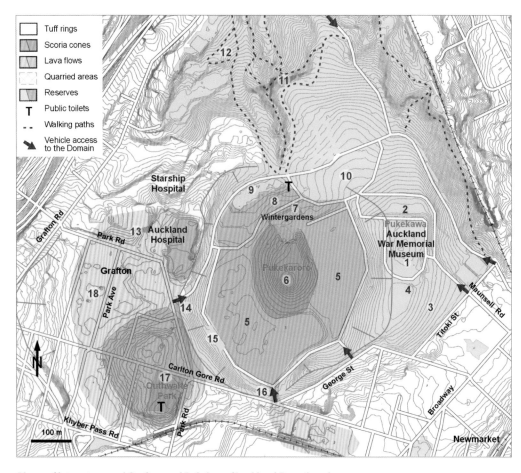

Places of interest around Grafton and Pukekawa/Auckland Domain volcanoes:

1. Auckland War Memorial Museum. Built on the eastern crest of the Domain tuff ring.
2. Auckland Cenotaph (1929) in front of museum commemorates those Aucklanders who died in the world wars. Views across Waitematā Harbour to Rangitoto, North Head and Mt Victoria volcanoes.
3. Ring roads parking. Site of transit housing soon after the Second World War.
4. Underground parking for museum, excavated out of tuff ring.
5. Basaltic lava-lake-filled explosion crater – 'moat' between central scoria cone and tuff ring.
6. Central scoria cone with Princess Te Puea Herangi's tōtara tree.
7. Wintergardens (built 1920s) with fernery behind in old scoria quarry; exposed scoria.
8. Domain kiosk, built as an Arts and Crafts Cottage for the Auckland Exhibition, 1913.
9. Duck ponds. Source of Auckland City's first piped water supply in 1865.
10. Historic bandstand, built 1912.
11. Cliffs of bedded tuff under forest, beside Lovers Lane walking track.
12. Cherry tree grove. Visit during flowering in early spring.
13. Auckland Hospital on Domain tuff ring, which here also buries the Grafton tuff ring and small scoria mound.

14. Art Deco Domain gates (1930s) on Park Rd, surmounted by a bronze nude male athlete.
15. Historic Domain cricket pavilion (grandstand), built 1898 to replace predecessor that was burnt down. Overlooks the sports fields on the floor of the western arm of the explosion crater. Site of many major events – first rugby league test in New Zealand, public receptions for royal visitors and the pope, Christmas in the Park.
16. Low saddle in tuff ring where lava from the crater lake spilled over during the eruption, possibly resulting in the lowering of the lava level in this part of the partially solidified crater lake. Several prefabricated buildings here are remnants from Camp Hale, the US barracks from the Second World War.
17. Outhwaite Park. Site of small scoria cone on the south side of Grafton Volcano.
18. Flat floor of Grafton crater infilled with basalt lava lake and mantled in ash from Domain eruptions.

◑ Dark brown scoria (right) from Outhwaite Park scoria cone (Grafton Volcano) overlain by light-coloured ash (left) of the Domain Volcano's tuff ring in a building excavation on Park Rd in 1998.

Pukekawa/
Auckland Domain

⊕ Auckland Domain Volcano from the south in 2009, showing its large explosion crater (with cricket wickets on the floor) and small vegetated central scoria cone. Auckland Museum (right) and Auckland Hospital (far left) are built on opposite sides of the tuff ring. *Photo by Alastair Jamieson*

Land status: Almost entirely in Auckland Domain, managed by Auckland Council.

What to do: Visit Auckland War Memorial Museum, and the tropical and cool houses at the Wintergardens and fernery; walk or run around the tuff ring and crater; see the sculptures or explore the bush; have a picnic, or play sport.

Geology

The 81-hectare Auckland Domain (established in 1845) sprawls over one of the older volcanoes in Auckland. The Domain Volcano appears to have erupted about 100,000 years ago through the eastern part of the fractionally older Grafton Volcano tuff ring.

Auckland Domain Volcano has a simple castle-and-moat layout with a small central scoria cone inside a large, shallow explosion crater with surrounding tuff ring. The museum is built on top of one side of the tuff ring and Auckland Hospital and the University of Auckland medical school on the other. The moat in the crater between the central scoria cone and the tuff ring is largely filled with a solidified lava lake, some of which may have overtopped a low point in the tuff ring at Carlton Gore Rd and flowed down

◔ Oblique aerial view of Auckland War Memorial Museum in 2018. It was constructed on the crest of the eastern side of the Domain volcano's tuff ring. *Photo by Alastair Jamieson*

towards Khyber Pass Rd. The lower level of the playing fields in the moat in front of the Domain grandstand may reflect withdrawal of some molten lava from beneath this part of the lava lake before it had solidified completely.

After eruptions ceased, the highly fractured basalt of the former lava lake became saturated with groundwater and a swamp developed on top of it, particularly in the lower area in front of the grandstand where up to 7 m of peat and clay accumulated over many thousands of years. In European times the swampy floor has been drained and covered with dirt to form playing fields on two levels.

◔ In autumn, brightly coloured anemone stinkhorn fungi can be seen growing amongst the leaf litter beside the path from the duck ponds down to Grafton Gully in Auckland Domain.

Human history

Auckland War Memorial Museum is appropriately sited on Pukekawa – 'hill of bitter memories', referring to the many local Māori killed on the isthmus in the 1820s musket wars. The museum was built in two halves – the front (opened in 1929) and back (opened in 1960) were constructed as memorials to all those citizens of the Auckland Province who lost their lives in the two world wars. The names of all 12,000 are inscribed on the walls of the two Halls of Memory on the top floor. Names above the windows commemorate battles from the First World War in the front building and from the Second World War around the back. During the Second World War, the Domain lawn in front of the museum was covered in American barracks, Camp Hale, 1942–44. After the war these buildings were shifted to behind the museum adjacent to Titoki St as transit housing for those waiting for a state house. During the Second World War, a further cluster of barracks was located in the southwest corner of the lower playing fields inside the Domain crater. Several of these are still present near the Carlton Gore Rd entrance.

The tuff ring has been modified in European times by the building of the museum, hospital and surrounding roads. The large, flat, windy area at the back of the museum that is popular amongst Auckland's kite-flying fraternity conceals one of Auckland's major water supply reservoirs (1952) and also the more recently constructed underground car park. Both have been buried in large excavations dug into the consolidated ash layers of the tuff ring.

The scoria cone, or castle, in the centre of the explosion crater is named Pukekaroro, or 'hill of the black-backed gull'. It retains the remnant earthworks of a former Māori pā. On its top is a tōtara tree planted by Princess Te Puea Herangi during Auckland's centennial celebrations (1940) and enclosed by traditional carvings and a mānuka palisade. An early quarry on the north side of the cone was transformed into the sunken fernery, behind the Wintergardens, in 1930.

⊙ A spindle-shaped volcanic bomb protrudes from the grass on the Domain scoria cone under the oaks.

Water supply
The Domain's duck ponds are freshwater springs, derived from groundwater flowing out of the fractured basalt and scoria that fill the crater. The source of this water is rain falling on the limited catchment of the explosion crater. Outpouring spring water has gradually eroded away the soft tuff ring on its northern side. Here a small stream cascades over numerous falls of bedded tuff towards Stanley St valley, which was a bay on the coast of the Waitematā Harbour when European colonists arrived. In 1865, with Auckland's population at 12,000, the city's first piped water supply was installed from the Domain ponds to the 'downtown' area. Not surprisingly the supply was inadequate and could not keep up with demand during dry summer periods. In 1877, the Domain water was replaced by a much more sustained source at Western Springs (see Three Kings Volcano on page 146).

◕ Barracks of the Second World War US Camp Hale in 1943 on the sloping lawn in front of the Auckland War Memorial Museum. *Courtesy of US Embassy and Consulate in New Zealand*

◕ The swampy floor of the Domain crater was drained and filled and opened as Auckland's premier cricket ground in 1874. *James D. Richardson, 1880s, Sir George Grey Special Collections, Auckland Libraries*
◕ The Auckland Exhibition was held in the Domain in 1913–14. This is the Palace of Industries located on the crater floor east of the present Wintergardens. *William T. Wilson, Auckland Museum*

Te Pou Hawaiki

● The quarried site of Te Pou Hawaiki scoria cone is occupied by a car park building (concrete block at bottom of photograph) for the University of Auckland's Faculty of Education in Epsom Ave, and is overlooked by Mt Eden to the north. *Photo by Alastair Jamieson, 2009*
● View from Mt Eden over the partly quarried-out Te Pou Hawaiki with Epsom Training College building behind, 1920s. *Photographer unknown, author's collection*
● This innocuous-looking sunken building with car parks on top is the Second World War bunker that was built in the side of the quarry that took away Te Pou Hawaiki. Inside the concrete is a two-storey underground building capped with 2 m of rock rubble. It was used as Auckland's civil defence headquarters for many years until the early 1970s. Photo 2018.

Land status: The site of Te Pou Hawaiki crater at 72 Epsom Ave is a car park building in the Epsom Campus of the University of Auckland, which allows public access.

Geology

Te Pou Hawaiki was the second-smallest cone in the Auckland Volcanic Field. Only Pukeiti was smaller. Both were typical steep-sided, small, welded scoria or spatter cones with bowl-shaped summit craters. Te Pou Hawaiki is located just south of Maungawhau and is surrounded by lava flows. It is thought to have erupted just before Mt Eden, approximately 28,000 years ago. Remote sensing studies have suggested that there may be a buried explosion crater and tuff ring beneath this site. This volcano could also have been the source of some of the surrounding lava flows that have been attributed to Mt Eden, as their chemistry is indistinguishable.

Human history

Te Pou Hawaiki means 'the pillar of Hawaiki', the ancestral homeland. This name indicates that the cone may well have been a place of considerable sanctity and ritual. It has no European name. Quarrying began on Te Pou Hawaiki's cone in the early 20th century and it was completely removed during the Second World War; excavation continued to about 15 m below ground level. More recently the disused quarry has been used for the construction of a three-storey car park building for the University of Auckland's Epsom Campus.

Maungawhau/
Mt Eden

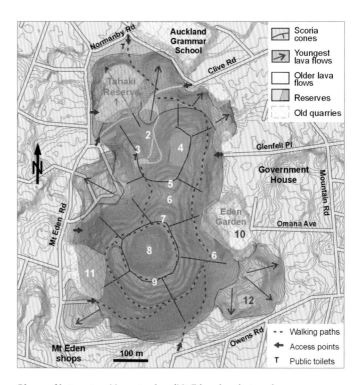

○ The top of Maungawhau/
Mt Eden from the southeast
showing the large crater
within the higher southern
scoria cone, with the
flattened top of the smaller
northern cone beyond. *Photo
by Alastair Jamieson, 2018*

**Places of interest on Maungawhau/Mt Eden showing scoria cones
and youngest lava flows:**

1. Tahaki Reserve. Face on south side of old quarry has scoria overlain by basaltic flow (from breached northern crater) and basalt feeder dike. Parking area, playground, stage.
2. Road up to kiosk rises up lava flow through breach in northern cone crater.
3. Historic kiosk (built 1927) and parking area. Walk up road to summit from here.
4. Summit of northern cone with spectacular view over central city; flat top of reservoirs buried in former north crater.
5. Highest point on crater rim of northern scoria cone with numerous pits from pre-European pā.

6. Extra-large pre-European kūmara pit.
7. Second-highest tihi (defended high point) in pre-European pā.
8. Te Kapua kai-a-Mataaho – 'the food bowl of Mataaho', Māori deity of volcanic activity = fountaining crater of southern cone.
9. Trig and highest point on cone. Former car park with panoramic views of the city.
10. Former quarry now Eden Garden and café (entry fee). Scoria, dikes feeding lava flows.
11. Granger's former scoria quarry, revegetated by the Friends of Maungawhau.
12. The Pines multi-storey residential block sits on one of Mt Eden's youngest lava flows.

Land status: The upper part of the cone is publicly accessible in reserve (Mt Eden and Tahaki reserves) administered by the Tūpuna Maunga Authority. The lower slopes of the cone are in reserves, Eden Garden, private properties and Government House grounds. Most of the lava-flow field that surrounds the cone is in suburban subdivisions with a scattering of public reserves like Withiel Thomas Reserve, Melville Park, Nicholson Park and Edenvale Park.

What to do: Take a round-trip walk up to the summit via the sealed road; admire the views and archaeology; have a picnic; take the children to the playground in Tahaki Reserve. Enjoy Eden Garden (entry fee) and café on Omana Ave.

Geology

Mt Eden Volcano is like a fried egg. It has a central scoria cone (yolk) surrounded by a field of lava flows (egg white). It is believed to have erupted through the lava flows of the earlier Mt St John volcano. The large, elongate scoria cone consists of two overlapping cones that erupted together or in close succession about 28,000 years ago. The crater of the lower, northern cone is now filled with two large buried water reservoirs. The highest point is on the southern crater rim of the younger scoria cone. The crater is the shape of an inverted, circular cone, about 180 m diameter and 50 m deep. Such a deep and perfectly shaped crater is rare in the Auckland Volcanic Field. In pre-European times this crater was known as Te Kapua kai-a-Mataaho – 'the food bowl of Mataaho', the deity responsible for volcanic activity.

Lava flowed out in all directions from around the base of the cone. The earlier, hotter and more fluid flows spread northeast to Khyber Pass Rd and Newmarket, southwest to Balmoral Rd, and west over the top of an existing flow from Mt St John to abut the slightly earlier flows from Three Kings near Gribblehirst Park in Morningside. Later flows were cooler and more viscous and did not travel as far from the base of the cone. These thick flows formed a steep-faced pedestal around

◗ View south over Maungawhau/ Mt Eden, 1937, showing the northern quarry (Tahaki Reserve) in the foreground before it was partially filled as a rubbish tip. Mt Eden Rd on right. *Whites Aviation Collection, Alexander Turnbull Library*

◐ Watercolour painting of Maungawhau/Mt Eden from the south, 1843. *John Guise Mitford, private collection, courtesy Alexander Turnbull Library*

the mountain, the edge of which runs close to Gillies Ave from Newmarket to Melville Park, then swings west at King George Ave, through the Windmill Rd reserve to Mt Eden Rd. Further north it runs through the grounds of Auckland Grammar and St Peter's College.

As the flows came out they cooled and solidified all the way back to their sources around the base of the scoria cone. Thus successively younger flows were expelled from vents higher and higher up the cone. The youngest flows were particularly cool and viscous and all squeezed out from vents 125 m ASL, forming sloping terraces on the lower slopes of the cone, like that on which The Pines apartment block has been built. One of these youngest flows breached the northern crater. This lava flowed north, creating the wide valley that is now the route of the access road up to the kiosk. Much of this young flow has been quarried away where Tahaki Reserve car park and stage are now located.

An unusual feature within one of Mt Eden's flows at the junction of present Edenvale and Wynyard roads (Edenvale Park) is a 5-metre-deep, 250-metre-long depression thought to be caused by the collapse of the lava-flow crust after the still molten lava beneath had flowed on and out. Within another Mt Eden flow in Mortimer Pass, Newmarket, is a pull-apart rift that has formed an unusual narrow cave with a slightly curved cross-section.

Pre-European Māori history

The Māori name, Maungawhau, means 'hill of the whau tree'. In pre-European times the top and sides of Maungawhau scoria cone were extensively modified to create a pā. On the outer slopes there are numerous flat terraces used partly for defence but mainly for living and working space. Many pits for crop storage have been dug on the flattened terraces. There was a series of three defended strong points on the highest scoria mounds on the crest, with the northernmost functioning as a communal food store with many pits. Unusually, there are no ditch and bank defences on this cone. Some of Maungawhau's earthworks have been lost or damaged by quarrying, and the construction of roads, parking areas and water reservoirs, but there is still much evidence of the former layout

of the pā. A marae area, Te maraeikohangia, was located in the northern crater (now reservoirs).

European history

The English name for this volcano was given by Governor William Hobson, founder of Auckland City, after his naval superior officer George Eden, Earl of Auckland. The trig station on the top of Mt Eden is the highest point (196 m ASL) on the Auckland Isthmus.

Several large quarries worked the thicker flows, operating for many years near Mt Eden Prison. During the 1840s and 1950s, dressed basalt stone was removed and used for some of Auckland's older stone buildings and for most of the 'bluestone' basalt kerb stones that line the city streets. Five other quarries were dug into the

flanks of Mt Eden's scoria cone prior to the 1920s. The excavated holes created by two of these are today the focus for two cherished reserves – Eden Garden and Tahaki Reserve, the latter partly infilled with refuse. The large quarry on the east side was developed from 1964 into today's world-class Eden Garden by a group of gardening enthusiasts. The other three quarries have left shallower scars on the steep western and southern slopes. The precise shape of the summit of the earlier, northern scoria cone is hard to determine because it has been modified by the construction of two buried reservoirs (in 1888 and 1912) that now provide an artificial flat top covered in grass.

The road to the top of Mt Eden was constructed by prison labour between 1869 and 1880. In 2016, it was closed to vehicles, except in special cases for the handicapped and elderly who are unable to walk up themselves. The upper parts of Mt Eden were gazetted a recreation reserve in 1876 and the Domain Board was established in 1879. Since the 1980s, a citizens' group of local volunteers, the Friends of Maungawhau, has been restoring native vegetation, improving tracks, advocating for improved management and celebrating its heritage in annual Love Your Maunga/Mountain days.

⊙ Profile of the eastern side of Mt Eden taken from near Manukau Rd in the 1860s. Note the extensive basalt stone walls and the forest-covered escarpment (vicinity of present-day Gillies Ave) around the viscous lava-flow pedestal. *John Kinder, Auckland Museum*
⊙ Te Kapua kai-a-Mataaho – the main crater of Maungawhau/Mt Eden was formed by fire-fountaining eruptions about 28,000 years ago.

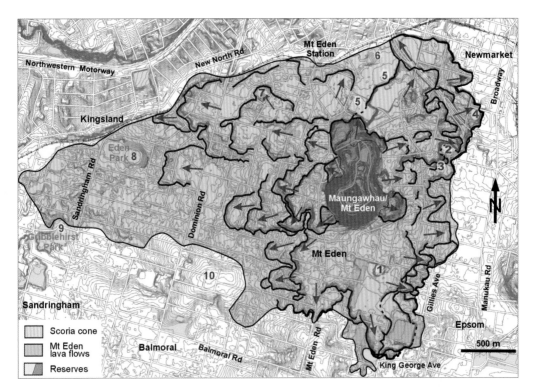

⊙ Map showing the extent of Maungawhau/Mt Eden's lava flows and the steep fronts and flow directions of younger flows.

Places of interest on the Maungawhau/Mt Eden lava-flow field:

1. Site of Te Pou Hawaiki volcano, quarry now filled with three-storey car park building. Second World War concrete bunker building to left of car park entrance.
2. Badminton Hall, 99 Gillies Ave. Old quarry with exposures of lava flows.
3. Withiel Thomas Reserve. Remnant of native forest growing on lava flows.
4. Unique Mortimer Pass lava cave. A fissure in the frontal lobe of a basaltic flow. Very narrow – do not enter.
5. Former Mt Eden Prison quarries in thick lava flow with large cooling columns. Used by rock climbers. In Auckland Grammar School grounds.

6. Mt Eden Prison. Older buildings built from blocks of Mt Eden basalt.
7. Edenvale Park. Depression formed by large collapsed lava cave.
8. Eden Park. Swampy depression between lava flows.
9. Gribblehirst Park. Depression between Mt Eden and Three Kings lava flows was a pond that accumulated diatomite.
10. Mont Le Grand Rd sandstone ridge, not overtopped by lava flows.

Ōhinerangi/Mt Hobson/ Ōhinerau

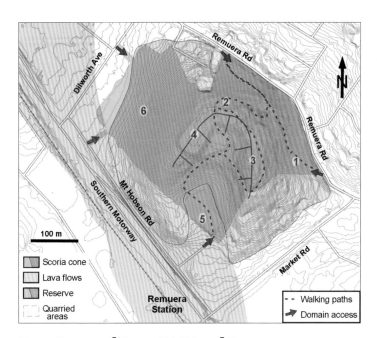

Places of interest on Ōhinerangi/Mt Hobson/Ōhinerau:

1. Main entrance path with daffodils that flower in early spring.
2. Seats for admiring the view over the Waitematā Harbour and beyond.
3. Defensive earthworks and storage pits from Ōhinerangi pā.
4. Flat top of water reservoir buried beneath crest of scoria cone.
5. Horseshoe-shaped crater breached to the southwest by lava flows. Flat floor is site of Second World War American mobile hospital store and a buried water reservoir (1955).
6. Northwestern grassy slopes facing the Newmarket Viaduct donated for reserve by the Dilworth Trust.

◔ View from the south over Mt Hobson and its U-shaped crater, breached to the southwest (foreground) by lava flows that flowed north and south down the route of the present Southern Motorway. On the crater floor, note the short grassed rectangle (buried reservoir). *Photo by Alastair Jamieson, 2018*

● The flat-topped profile of Ōhinerangi/Mt Hobson viewed from the northwest is created by the grassed-over roof of a water reservoir installed on the crest in 1935. The grassed western slopes facing the motorway (on right) are artificially smooth because the terraces, once carved into the hillside by pre-European Māori, were buried by rubble from the reservoir excavations.

Land status: The upper part of the cone is publicly accessible in reserve administered by the Tūpuna Maunga Authority. The lava flows and some of the lower slopes of the cone are beneath private properties.

What to do: Walk up a track to explore the fabulous views and archaeology on the summit; admire the display of daffodils around the main entrance in early spring.

Geology

Mt Hobson is a moderately large scoria cone that erupted through the Remuera Rd sandstone ridge approximately 35,000 years ago. Fountaining eruptions built up the cone (143 m ASL) around a single vent. The central crater was breached to the southwest by a small lava flow, creating a horseshoe-shaped depression. The longer northern tongue of this lava flow reached almost as far as the Newmarket Railway Station. Exposures of the basalt lava flow can be seen alongside the railway line and under the Market Rd bridge over the motorway where it is buried beneath volcanic ash erupted later from Three Kings.

Human history

The ancient Māori name for this volcano is Ōhinerangi, meaning 'the dwelling place of Hinerangi', the goddess of whirlwinds. Later Māori names applied to this cone are Ōhinerau = 'dwelling place of Hinerau' and Remuera

(corruption of Remu-wera), meaning 'the burnt hem of a garment', from the 1700s when a young Hauraki woman was killed and eaten by the Waiohua inhabitants of the pā near present-day Dilworth School senior campus.

Many of the terraces, ditch defences, storage pits and middens from a pre-European pā are still clearly visible around the cone's crest, except where reservoir construction has destroyed them. The pā had a single strong point, or tihi, on the highest point, defended by two large terraces and terrace scarps. Mt Hobson's western slopes, seen from the Newmarket Viaduct, are thought to have been extensive hillside kūmara gardens.

The English name Mt Hobson was given in 1841 to honour Governor Hobson, New Zealand's first governor general and European founder of the city of Auckland. It was the first volcanic cone climbed by Hobson after his arrival on the isthmus.

In 1877, a public debate raged over the need to sacrifice this prominent landscape feature to build

◔ The eastern side of Mt Hobson viewed from the slopes of Mt St John. *Hugh Boscawen, 1899, Auckland Museum*
◑ Eastern side of Mt Hobson in 1958 before the Southern Motorway had been constructed between the railway line and Mt Hobson Rd in the foreground. Hobson Bay is in the top left. *Whites Aviation Collection, Alexander Turnbull Library*

the roads of the growing city. Good sense prevailed and the 9.5-hectare Mt Hobson Domain was gazetted in 1880. Scoria was excavated from two small quarries on private land on the northern side between 1914 and 1928. One is now the site of a private tennis court. Other modifications to the scoria cone include houses built high up on the northern (Remuera Rd) and southeastern (Market Rd) sides, and two buried reservoirs – one on the crest (1900, replaced in 1935 and 2017) and one in the floor of the crater (1955). In 1921, the Dilworth Trust donated 5 hectares on the

western slopes of Mt Hobson to be added to the reserve and this forms the grassy slopes so prominently visible from the Newmarket Viaduct.

In 1942–44, during the Second World War, a large medical store for the US Naval Hospital in nearby Hobson Park (now Dilworth School) was located on the floor of the crater. The grassed slopes on the north side above Remuera Rd turn golden in early spring with the flowering of masses of daffodils, gifted by residents of Remuera to commemorate all those who gave their lives in the Second World War.

Te Kōpuke/Tītīkōpuke/ Mt St John

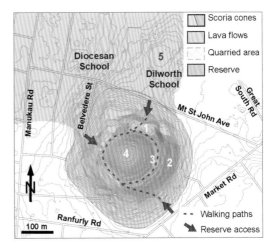

Places of interest on Te Kōpuke/Tītīkōpuke/Mt St John:

1. Track to summit passes through small 19th-century scoria quarry.
2. Crest and sides of cone have extensive Māori defensive earthworks and dwelling sites.
3. Concrete bases of two Second World War anti-aircraft gun batteries in pits.
4. Crater floor underlain by 4 m of sediment which has accumulated in the pond that forms after heavy rain.
5. Dilworth School buildings built on crest of lava flow from Mt St John.

Land status: The upper part of the cone is publicly accessible in reserve administered by the Tūpuna Maunga Authority. The lower slopes are covered in private dwellings.

What to do: Take a walking track to the summit and around the crest of the crater; enjoy the views and archaeology.

⊙ The simple scoria cone and crater of Mt St John viewed from the southeast in 2009. Lava flowed out from the western (left) side of the cone, possibly rafting some scoria away and leaving that side lower. Note the pre-European Māori pits and terraces on the crest and slopes of the cone. *Photo by Alastair Jamieson*

129

◉ A temporary pond forms in the ash-lined crater of Mt St John after heavy rain in 2014.

Geology

Mt St John is a small scoria cone, 126 m ASL, that was formed by dry fountaining eruptions approximately 75,000 years ago. Its cone is thinly blanketed with ash thrown out from Three Kings 28,500 years ago. This ash has lined the inside of the crater, which, as a result, retains a pond after heavy rain. This pond has accumulated 3 m of clay overlain by 0.6 m of swampy peat loam. The latter contains fossil spores and pollens that reveal the changing vegetation on Mt St John over the last 10,000 years.

Lava flowed out from the base of the scoria cone on the western side and down a valley to the west forming Auckland's longest flow (see Meola Reef Te Tokaroa on pages 131–33). Ferdinand von Hochstetter, who mapped this area in 1859, shows a small mound of scoria on the site of the present-day Dilworth senior school buildings. This scoria may have been rafted there by a small lava flow from neighbouring Mt St John.

Human history

The Māori name, Te Kōpuke, means 'the prominent mound'. The crest of the cone has numerous food-storage pits and terraces from the pre-European pā that once presided over a renowned gardening area. The flattened crest of the cone is divided into several defensive compartments by three transverse ditches and associated banks.

This cone was given its English name in 1840 after St John the Evangelist, patron saint of the clerics in Auckland. In 1872, 4 hectares around the summit of Mt St John was declared an education reserve and vested in the Crown as a public recreation ground in 1896. There have been a number of additions over the years. In 1978, it was gazetted for protection as a recreation reserve and historic place (archaeological). The site of a small 19th-century quarry is recognisable on the upper northern slopes.

◉ Te Kōpuke/Mt St John from Mt Hobson in 1909. Note the overgrown quarry high on the northern side of the scoria cone (right). *Sydney Charles Smith, Alexander Turnbull Library*

⊙ Watercolour painting of a Paremata 'Māori feast in return for favours' hosted by Waikato chiefs Te Wherowhero and Wetere Te Kauae, on 11 May 1844 near the junction of present Market and Great South roads, Epsom. The prominent cone is Te Kōpuke/Mt St John (erroneously identified by most historians as Mt Hobson), with horizon profiles of Mt Eden (right) and Three Kings cones (left). *Joseph Merrett, Hocken Collections, University of Otago*

Meola Reef Te Tokaroa

Land status: The seaward 2 km of the linear reef is accessible when the tide is low; the landward end is Meola Reef Reserve Te Tokaroa managed by Auckland Council.

What to do: Take the leisurely 1.5 km loop track around the reserve; venture out onto the reef from the reserve or at low tide walk across sand flats from the end of Garnet Rd to the reef; exercise your dog in the off-leash dog area in the reserve. Inland and still on this lava flow you can visit MOTAT, Auckland Zoo and Western Springs (see page 318).

Geology

In 2008, University of Auckland graduate student Jeremy Eade used geochemical analyses to identify lava that had erupted from Mt St John, Mt Eden, One Tree Hill and Three Kings. A surprising outcome was that the lava flow that forms Meola Reef, between Pt Chevalier and Herne Bay, has the same chemistry as Mt St John and could be tracked back beneath some of the lava flows from Three Kings and Mt Eden. We now know that Mt St John erupted 75,000 years ago, well before Mt Eden (28,000 years ago), and that a large lava flow poured from Mt St John's base and flowed down a valley that passed beneath present-day Mt Eden and westward to join the Waitematā River valley near Birkenhead. This river of lava cooled to form a ribbon of solid basalt and Auckland's longest lava flow – 11 km from source to toe.

Te Kōpuke erupted when the climate was much cooler than now, sea level was lower and the Waitematā Harbour was a forested valley. The lava flow filled the valley it flowed down and displaced much of the stream water, which began flowing along both sides and eroding the softer rocks of the former valley sides. Today these are Motions and Meola creeks – one on either side of the Mt St John lava flow. Some water also flows down the original valley course through the fractures and broken-up blocks in the lava flow

131

and appears at the surface as it nears tide waters as Western Springs, which feeds Motions Creek.

Human history

An early Māori name for Meola Reef is Te Ara whakapekapeka-o-Ruarangi, meaning 'the pathway or diversion of Ruarangi', as the reef is said to have been used by Ruarangi to cross the Waitematā Harbour as he fled from his brother Oho Matakamokamo. Other names for Meola Reef are Te Tokaroa (the long reef), Westmere Reef, Pt Chev Reef and Black Reef. The name Meola is thought to have come from a glacier in India near where an early colonial resident of Alberton, Allan Kerr Taylor, was born.

After about 1900, suburbia started to creep out around both sides of the bay containing Meola Reef, although the publicly owned rocky land between Western Springs and the reef remained undeveloped. During 1960–76, Meola Reef Reserve was a major Auckland refuse tip on top of the lava flow. Subsequently it was covered with soil and grassed for cattle grazing and recreational reserve. Native tree plantings around the perimeter of the reserve began in the 1990s.

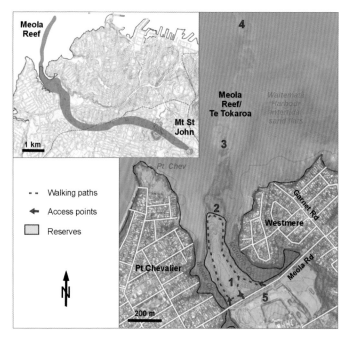

⊙ Map showing the route of the longest lava flow in Auckland, which poured out from the base of Mt St John and flowed 11 km to the ancient Waitematā River valley near Birkenhead. Also shown are places of interest around Meola Reef at the seaward end of the flow.

Places of interest at Meola Reef Te Tokaroa:

1. Meola Reef Reserve has been built up above sea level by fill from a former rubbish tip.
2. Step off walkway and venture onto the lava flow at high-tide level. See the tops of hexagonal and pentagonal basalt cooling columns in the lava flow. Mangroves and salt marsh plants grow out of cracks in the lava flow.
3. At mid-tide most of the flow is covered in dense Pacific oysters, accidentally introduced to Auckland from Japan in the 1960s. They are razor-sharp so be extremely careful.
4. At spring low tide, visit the seaward end of Meola Reef via the sand flats from Garnet Rd. View bright orange, yellow and purple sponges among brown seaweed and colourful sea squirts, chitons and sea slugs under low-tide rocks. Gently replace the rocks.
5. **MOTAT Aviation Workshop and displays.**

◔ View northwest over
the landward end of Meola
Reef in 1963 just as landfill
operations were about
to start. *Whites Aviation
Collection, Alexander
Turnbull Library*

◑ The lava flow that
forms Meola Reef Te
Tokaroa extends out into
the Waitematā Harbour
almost reaching across to
Birkenhead. Meola Reef
Reserve is the grassed area
on top of the flow at the
head of the bay. *Photo by
Alastair Jamieson, 2009*

133

Maungakiekie/
One Tree Hill

◔ One Tree Hill scoria cone with its one intact and two breached craters viewed from the southwest. *Photo by Alastair Jamieson, 2018*
◑ One Tree Hill profile from Mt Eden, 2008.

Land status: The scoria cone and upper lava flows are partly in One Tree Hill Domain, administered by the Tūpuna Maunga Authority, and partly in Cornwall Park, administered by the Cornwall Park Trust. The extensive lava-flow fields are mostly beneath private properties and numerous public roads, with some in reserves.

What to do: The most popular activity is walking around or up the cone, but there are many alternative walks, especially in summer when the paddocks are dry. Jogging is also popular; or go for a coffee or a picnic, or take the children to the playground.

Geology

Maungakiekie is the second-largest volcano in the Auckland Volcanic Field with 2000 hectares of lava-flow fields surrounding a large, complex scoria cone. The highest part of the cone (183 m ASL) is made of scoria that fountained out of the adjacent crater as frothy lava and built up on its northeastern side during a period of southwesterly winds. Two other craters in the southwest and southeast are horseshoe-shaped, having been breached by lava flows that carried away the scoria ramparts on their downhill sides.

Lava also burst through the northern lower flanks of the cone and the resulting lava flows spread out as a broad apron completely surrounding the volcano. To the north the flows may have reached Khyber Pass Rd in Newmarket. An extensive fan of lava spread 3 km south to the Onehunga shoreline of the Manukau Harbour. The eastern edge of lava flows from One Tree Hill runs beneath Ellerslie township and northwards to Remuera Intermediate School and Ascot Hospital in Greenlane Rd. The flow front blocked a shallow valley in the vicinity of Ellerslie Racecourse, creating a low-lying swamp between the lava flows and the existing sandstone ridges of Remuera Rd and Ladies Mile. The swamp was drained in European times and became the premier jockey racing club in New Zealand. About 1.5 km east of the One Tree Hill cone, the lava flowed right around a sandstone hill at Oranga but was not deep enough to completely engulf it.

Like Maungawhau, the youngest flows were extruded as cooler, more viscous lava than the earlier ones and from a higher elevation on the outer slopes of the scoria cone. These later viscous flows cooled and solidified without advancing far from their vents and form the terrace and spur on either side of Huia Lodge and the kiosk today.

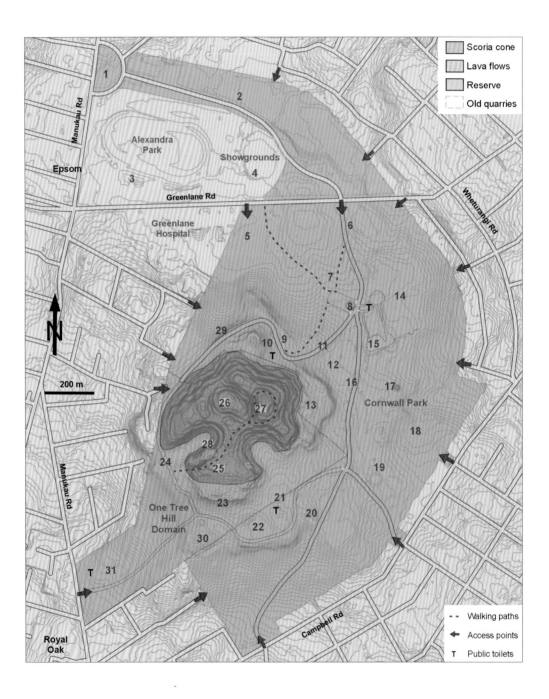

Scoria cone
Lava flows
Reserve
Old quarries

Manukau Rd

1

2

Alexandra
Park

Showgrounds

Epsom

3

4

Greenlane Rd

Wheturangi Rd

Greenlane
Hospital

5

6

7

14

8   T

N

29

10   9

11

15

200 m

12

16

17

26   27

13

Cornwall Park

28

18

24   25

19

Manukau Rd

23

21   T

20

One Tree
Hill
Domain

22

30

31   T

Campbell Rd

Royal
Oak

- - Walking paths
← Access points
T   Public toilets

Places of interest on and around Maungakiekie/One Tree Hill scoria cones:

1. Fountain and John Logan Campbell statue (opened 1906).
2. Puriri Drive with rugby and rugby league fields on north side.
3. Alexandra Park raceway (home of Auckland Trotting Club).
4. ASB Showgrounds events centre, Epsom. Venue for Auckland's Easter Show.
5. Cornwall Cricket Club ground and tree-lined former grand avenue entrance to the park.
6. Pohutukawa Drive main entrance to Cornwall Park, off Greenlane Rd.
7. Cherry tree grove, very popular for picnics when flowering in September.
8. Roundabout and sunken gardens. This is artificial and not a natural landform or crater.
9. Memorial steps (1956) for Sir John Logan Campbell, donor of Cornwall Park.
10. Huia Lodge information centre and displays, Acacia Cottage, tea kiosk restaurant and café.
11. Road leading up to kiosk (Michael Horton Drive) runs along crest of one of the youngest lava flows from the cone.
12. Groves of kauri, tōtara and rimu planted in the 1940s.
13. Terrace on the lower northeastern slopes of the cone was formed by viscous lava flow that was among last to erupt from the volcano and did not travel far.
14. Café. Site of US Army Hospital, later Cornwall Hospital. Toka Tu Whenua – the rock pillar (rongo stone) that held the mauri, or spiritual essence, of Te Tātua-a-Riukiuta/Three Kings – now stands near the band rotunda and magnolia grove.
15. Parking area with barbecues provided.
16. Twin Oaks Drive. Trees planted in 1934. Extremely popular with walkers when thousands of daffodils and jonquils are flowering in early spring.
17. Folly. Artificial conical mound of earth; not a natural volcanic feature.
18. Exposures of lava-flow basalt in paddocks with a few ropey pahoehoe surfaces.
19. Cornwall Park farm buildings, woolshed and yards. No public access.
20. The Grotto fernery and karaka grove; an old quarry where lava-flow basalt can be seen.
21. Basalt toilet block and Auckland Archery Club at mouth of breached southern crater.
22. Flat surface in paddock is top of buried water reservoir.
23. Bracing blocks and remnants of a Second World War RNZAF signal base.
24. Parking area at mouth of breached western crater; for those walking up to the summit.
25. Road up to summit, now closed to vehicles. Follows main entry into Maungakiekie pā.
26. Main unbreached crater; walk around circular crest with two water reservoirs, one buried.
27. Summit of Maungakiekie. Panoramic views of Auckland. Obelisk (unveiled 1948; 33 m high) and grave of Sir John Logan Campbell.
28. See scoria faces in disused 1850s quarry.
29. Old olive grove planted on Campbell's farm. Road cuttings of basalt from one of youngest and most viscous lava flows that formed the spur between here and Huia Lodge.
30. The Sorrento Restaurant in One Tree Hill Domain was built as the clubhouse for the Maungakiekie Golf Club.
31. Stardome Observatory and Planetarium (open daily), children's playground and flying fox.

● Auckland Golf Club (1901–23) and later Maungakiekie Golf Club (1923–42) were located on land in One Tree Hill Domain and Cornwall Park. Three Kings tuff ring and scoria cone in the top right. *Auckland Weekly News, 23 Sept 1909, Sir George Grey Special Collections, Auckland City Libraries*

● Sir John Logan Campbell's coffin makes its journey up the recently constructed access road to the summit of Maungakiekie for the burial ceremony in 1912. *Sir George Grey Special Collections, Auckland City Libraries*

● View west over One Tree Hill Domain and Cornwall Park in 1947 showing the extent of Cornwall Hospital in the foreground, constructed during the Second World War as a US Army hospital. It was demolished in 1975. *Whites Aviation, University of Auckland*

⊙ Cornwall Park is home to the most spectacular cherry blossom display in Auckland and attracts crowds of picnickers for a few weeks each spring.

One Tree Hill erupted about 60,000 years ago. Its lava-flow fields were later pierced and partly buried by scoria cones and flows erupted by Mt Eden, Mt Smart and Te Hopua volcanoes. Most of the lava flows in One Tree Hill Domain, Cornwall Park, Royal Oak and Epsom have been buried by a smooth 2–10-metre-thick blanket of ash erupted 28,500 years ago from the Three Kings Volcano.

Pre-European Māori history

Maungakiekie is the Māori name for the cone, meaning 'the hill of the kiekie vine', although today the only kiekie growing in the park is planted next to Huia Lodge. Maungakiekie is the most extensively terraced of all Auckland's volcanic cones and one of the largest pre-European archaeological site complexes in New Zealand. It includes dozens of house sites and garden terraces, and numerous groups of food-storage pits. The mountain lay at the centre of Ngā Māra-o-Tahuri – the expansive 'gardens or cultivations of Tahuri', a Waiohua ancestress. This scoria cone pā is one of the largest pre-iron-age forts in the world. Its four summits were all heavily defended by ditches and wooden palisades. On its peak was the tihi, the most sacred and heavily defended part of the complex. It was named Te Tōtara-i-ahua after the sacred lone tōtara that, according to legend, grew there in pre-European times.

Cornwall Park

One Tree Hill was purchased from Māori by Thomas Henry in 1845 and developed into his 400-hectare Mt Prospect Estate. Governor Grey disallowed the purchase of the hilltop and the southwestern slopes, setting them aside as a reserve in 1848 and this became One Tree Hill Domain. The rest was purchased by William Brown and John Logan Campbell in 1853 and owned solely by Campbell from 1873. He gifted the park (named in honour of the visiting Duke and Duchess of Cornwall) to the people of New Zealand in 1901. Open for public viewing is Acacia Cottage (built 1841), Auckland's oldest house and Campbell's original home, which was moved to the northern slopes of One Tree Hill in 1920 from behind Campbell and Brown's store in Shortland St in the inner city. Opposite it is the restored Huia Lodge Visitor Centre, built as a caretaker's cottage in 1903.

In the 1860s, Campbell began tree planting, including an olive grove on the northern side with the intention of starting an oil industry. He purchased the best grass for pasture and well-bred stock for breeding. Extensive drystone walls were constructed between the paddocks and alongside Greenlane Rd using basalt

collected from the surface of the surrounding lava-flow field. Campbell died in 1912 and was buried on the summit of Maungakiekie. The 21-metre-tall obelisk beside his grave was bequeathed by him as a memorial to the Māori people of 1840. It is made of basalt and Coromandel Granite and was finally completed in 1948 and unveiled by the Māori King, Koroki. The summit road was opened in 1907 and closed to vehicles in 2018.

During the Second World War, 26 hectares on the northeastern boundary of Cornwall Park were used for the construction of a sprawling, single-storey, 1500-bed US Army general hospital. After the war, Cornwall Hospital was used as the National Women's Hospital until 1964 and then for geriatric patients until 1973. It was demolished in 1975 and the land rehabilitated to grassed parkland with a barbecue and picnic area, band rotunda and, recently, a café. Today, the grass in much of One Tree Hill Domain and parts of Cornwall Park is kept short by the grazing of a small herd of about 100 cattle and a flock of

◑ This bronze statue of Sir John Logan Campbell and fountain at the Manukau Rd entrance to Cornwall Park was erected in 1906 to mark the former mayor of Auckland's generosity in gifting the park to the people of New Zealand.

❸ One Tree Hill profile
from Mt Hobson in the
north in the mid-1870s.
*John Kinder, Auckland
Public Art Gallery*

around 800 Romney sheep. In spring, a major
attraction in Auckland is to walk through the
daffodils in Twin Oaks Drive, picnic under the
flowering cherry trees off Pohutukawa Drive and
view the newborn lambs in Cornwall Park. Land
leased from Cornwall Park has become the home
of numerous sports clubs, including golf, tennis,
rugby league, rugby, bowls, cricket, archery and
harness racing.

Summit tree

A tōtara tree known as 'Te Tōtara-i-ahua' was
planted on the summit of Maungakiekie in the
early 17th century, but by the time European
colonists arrived it was no longer there and had
been replaced by a large pōhutukawa. It was
from this pōhutukawa that the volcano got its
European name One Tree Hill, but it was felled
by an unknown colonist in 1852. The felled tree
was replaced in the 1870s by Sir John Logan
Campbell with a group of Monterey pines,
of which only one survived through to 2000,
when it was cut down after being attacked by

a chainsaw-wielding Māori activist. A grove of
nine young tōtara and pōhutukawa was planted
on the summit in June 2016 with the intention
of eventually thinning them down to the single
strongest tree.

Water supply

Rain falling on the southern slopes of One
Tree Hill's extensive lava-flow field (from
Royal Oak to Te Papapa) percolates into the
underlying lava flows and on down through the
broken rock towards the Onehunga foreshore
of the Manukau Harbour. As this groundwater
reaches sea level it encounters denser salt
water that has penetrated landward within
the basalt. The fresh water flows over the top
of the salt and bubbles up at the surface in a
number of places close to the old shoreline.
Today, most of these springs, except Captain
Springs, are hidden beneath buildings.

Water from the largest of these Onehunga
Springs was used from 1854 to power a water-
wheel at Bycroft's Flour Mill. A second spring

◐ Several large pōhutukawa beside the Cornwall Park kiosk have unusual yellow flowers early in summer. This rare form originates from Mōtītī Island in the Bay of Plenty.
◑ Map showing the extent of Maungakiekie/One Tree Hill Volcano – the second largest in the field after Rangitoto. The exact boundaries of One Tree Hill's lava flows are unknown where they are buried beneath the later lava flows of Mt Smart in the southeast and Mt Eden in the northwest.

nearby was tapped for water to supply One Tree Hill Borough from 1891 to 1948. Water from both springs was pumped to reservoirs on One Tree Hill scoria cone to provide the gravity head for reticulation. At various times this Onehunga groundwater was also supplied to households in nearby Epsom, Ellerslie, Mt Roskill and Māngere Bridge. These springs have supplied Onehunga with piped water since 1878 and it still feeds into the Auckland water supply. Today, Onehunga water is unfluoridated by request of the locals and fed just to them. It can be interconnected with the main Auckland supply and excess supplied to the whole system. There are currently three water reservoirs on One Tree Hill, two on the western crest of the scoria cone and one on the upper lava flows in the southern part of Cornwall Park.

◐ *Places of interest on Maungakiekie/One Tree Hill lava-flow field:*

1. One Tree Hill lava flows may extend as far north as Newmarket.
2. Ellerslie Racecourse occupies former swamp dammed between lava flows and sandstone ridge that underlies Ladies Mile.
3. Lava flows flowed around sandstone hill at Oranga.
4. Liverpool St exposure of Three Kings tuff.
5. Uneven surface of One Tree Hill lava flows buried by Three Kings tuff formed shallow swampy hollows (now drained) that Hochstetter mistook for volcanic explosion craters.
6. Seymour Park and surrounds are a former swamp dammed between lava flows and sandstone ridges to south and west (Monte Cecilia Park).
7. Te Hopua Volcano erupted through One Tree Hill lava flows.
8. Toes of basalt lava flows form parts of Manukau Harbour shoreline near Manukau Cruising Club.
9. Toes of basalt lava flows form parts of Manukau Harbour shoreline near Waikaraka Cemetery.
10. Bycroft Reserve (former springs) where water is pumped into Auckland water supply from One Tree Hill lava flows.
11. Captain Springs Reserve where water flows up from cracks in One Tree Hill lava flows.
12. Hochstetter Pond and Puka St Grotto are collapsed portions of a lava cave.
13. Namata Rd cuttings through basaltic lava flows.

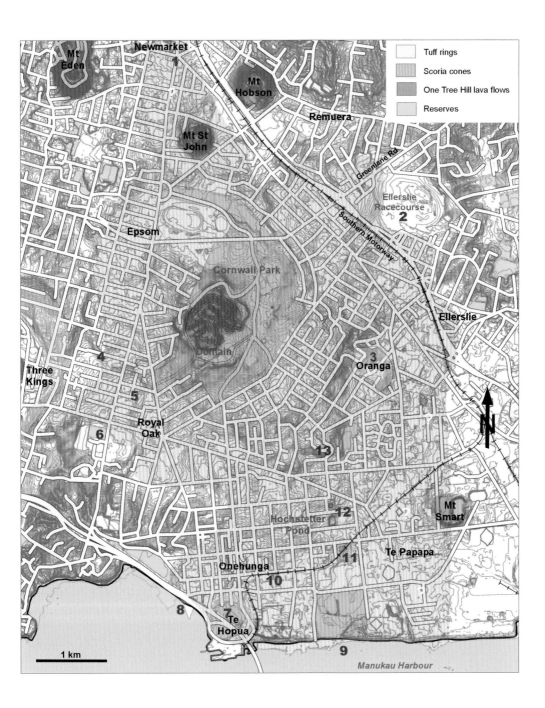

Tuff rings
Scoria cones
One Tree Hill lava flows
Reserves

Mt
Eden

Newmarket

Mt
Hobson

Remuera

Mt St
John

Greenlane Rd

Ellerslie
Racecourse
2

Epsom

Southern Motorway

Cornwall Park

Ellerslie

The
Domain

4

3
Oranga

Three
Kings

5

13

Royal
Oak

6

N

Mt
Smart

12

Hochstetter
Pond

Te Papapa

11

8

Onehunga

10

7
Te
Hopua

9

1 km

Manukau Harbour

Hochstetter Pond and Puka St Grotto

Land status: Hochstetter Pond is in a public reserve, whereas Puka St Grotto is on private land and you need permission from the owners to view it.

Where they are: Hochstetter Pond is at 36 Grotto St and Puka St Grotto is at 5 Puka St, Onehunga.

In Onehunga there are two large depressions amongst the houses that were formed by collapse of the roof of a large lava cave. The collapses probably occurred while hot lava was still flowing through the lava tube inside the otherwise cooled and solidified basaltic flow from One Tree Hill Volcano. The pieces of cave roof that collapsed were probably mostly carried away through the tube by the moving lava and the holes were progressively enlarged by this process until the lava stopped flowing. These two depressions were first noticed by Charles Heaphy and Ferdinand von Hochstetter in the 1850s. Heaphy thought they were small volcanic craters whereas Hochstetter argued that they were formed by collapse of sections of roof of a large lava cave.

Ponds do not usually form within lava cave collapses as water easily drains through the cracks in the surrounding lava-flow basalt. In this instance it would appear that these depressions

were later draped in volcanic ash erupted from nearby Mt Smart, which could have sealed the Grotto St depression and allowed a pond to form. The 100-metre-diameter Grotto St depression is partly filled with up to 20 m depth of sediment. Upper parts of the sediment fill are diatomite, a white rock composed of the silica skeletons of microscopic algal phytoplankton (diatoms) that once lived in the open water of the pond. In the 1940s–50s, the pond was drained and some of the surface diatomite was mined for use in a locally marketed abrasive cleaner called 'Grotto Maid'.

In 2007, Auckland City Council purchased the property in Grotto St containing the pond to create a suburban wetland reserve. The property has had a chequered history since its public acquisition by Sir George Grey in 1855 to create a reserve, its later disposal by the government to a private buyer and more recently its scheduling for protection as a site of geological importance on local body district schemes. Uphill of Hochstetter Pond is a 50-metre-diameter, 10–12-metre-deep, vertically walled depression known as Puka St Grotto. It lies in the back sections of three privately owned properties.

◑ Hochstetter Pond, Grotto St, Onehunga. *Photo by Hugh Grenfell, 2008*
◑ Hochstetter's sketched cross-section showing the pond and grotto at Onehunga to be lava cave collapse features.
◑ Aerial view from the south in 2009 over Hochstetter Pond (in foreground) and Puka St Grotto (beyond), Onehunga, both formed by roof collapses in a thick One Tree Hill lava flow.

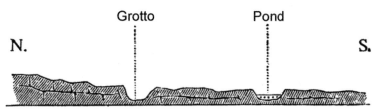

Te Tātua-a-Riukiuta/
Three Kings

○ View south over the site of Three Kings Volcano showing the extent of the large explosion crater and surrounding, house-covered tuff ring (left and right edges of photo). The circular suburban streets outline the shape of the crater. Big King with its summit water reservoir (centre right) is the only scoria cone left that has not been quarried away.
Photo by Alastair Jamieson, 2018

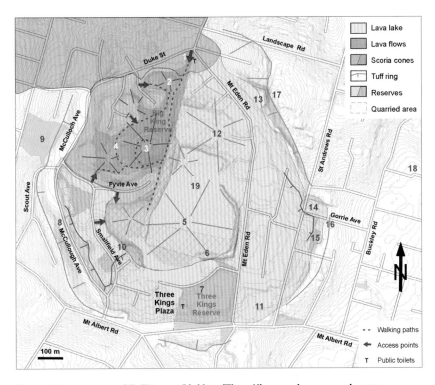

Places of interest around Te Tātua-a-Riukiuta/Three Kings scoria cones and crater:

1. Car park and main track entrance to Big King Reserve. Located where lava breached the tuff ring and flowed out as voluminous flows down to Western Springs.
2. The only one of the smaller Three Kings scoria mounds not quarried away.
3. Summit of Big King with water reservoir and views.
4. Pre-European kūmara pits and buried water reservoir.
5. Site of quarried-away South King/Taurangi.
6. Face of former quarry showing basaltic lava lake solidified against lower scoria slopes of South King.
7. Natural flat floor of partly drained lava lake; former pumphouse now used by Auckland Brass Band.
8. Site of former Wesleyan Native School in McCullough Ave 'moat' between tuff ring crest (west side) and crater lava-flow front (Smallfield Ave).
9. Arthur Richards Memorial Park on crest of low tuff ring of Three Kings Volcano.
10. View point over quarried-away scoria cones.
11. Three Kings School with high ground formed by crust of former lava lake.
12. Site of quarried-away East King/Koheranui.
13. Cliff exposure of bedded tuff of Three Kings tuff ring beside car park in Eden Mews.
14. Terrace on St Andrews Rd on top of solid edge of Three Kings crater lava lake.
15. Rowan Reserve has basaltic lava that solidified around the upper edge of the lava lake.
16. Rest of Three Kings tuff ring between 12 and 13 Gorrie Ave.
17. Highest part of Three Kings tuff ring (127 m ASL) near 187 St Andrews Rd.
18. Folded Three Kings tuff in road cutting outside 27 Liverpool St (see page 151).
19. New housing subdivision in partly infilled old quarry.

147

Land status: Most of the Three Kings crater and scoria cones are privately owned commercial or residential property or the former quarry, which is being rehabilitated. Besides the roads there are a number of small publicly accessible reserves, including the most significant, Big King Reserve.

What to do: Walk to the summit of remaining Big King for the view and take some of the other walks within this and other nearby reserves.

Geology

Three Kings was probably the most complex of Auckland's volcanoes. It erupted through the main sandstone divide between the Manukau and Waitematā catchments 28,500 years ago. The initial eruptions were massive wet explosive blasts that resulted in the largest explosion crater in Auckland – 1 km across and over 150 m deep. Volcanic ash was blown high into the air and spread northeast by the wind as far as Remuera. Ash deposits mantled the landscape – more thickly near the vent, thinning and fining further away. Close to the crater, base surges

of searing hot ash, augmented by larger debris, blasted outwards, building up a tuff ring that completely encircled it. Today the tuff ring crest is roughly defined by Mt Albert, St Andrews and Landscape roads, Duke St, and Scout and Simmonds avenues. The highest part of the tuff ring built up on top of an existing sandstone ridge forming the high point at 126 m ASL near the junction of St Andrews and Landscape roads.

Following the wet explosive eruptions, dry fountaining began from a number of different vents within the existing crater. Over a few weeks, the crater was partly filled with scoria

◑ Panorama over Three Kings scoria cones from the northeastern crest of the tuff ring, 1868. From the left the cones are East King, South King and West (Big) King. *Richardson Album v.5, Auckland Museum*
◐ View over the remains of Three Kings scoria cones from the northeastern crest of the tuff ring (St Andrews Rd), 2010. Same view as 1868 photo. Mt Eden Rd is in the foreground.

and three significant scoria cones piled up to
45 m above the new floor level. A dozen or so
smaller scoria mounds, each more than 10 m high,
were also produced. Fountaining eruptions were
unusually powerful, with fine, pea-sized scoria
lapilli being blown northeastwards more than
a kilometre beyond the crest of the tuff ring.

As the fountaining was building up the
scoria cones, molten lava burst forth from their
base and filled much of the crater between the
cones and the arcuate southern half of the tuff
ring. The lava was cooler and more viscous in
the southwest where its front solidified before
reaching the crater's inner wall. A deep, curved
moat remained between this lava-flow front
and the tuff ring. Today, McCullough Ave runs
along the floor of this moat and Smallfield
Ave runs along the crest of the flow front.

In the southeast, the lake of molten lava
filled the gap between the scoria cones and the
tuff ring up to the level of the tuff ring crest, in
the vicinity of present-day Three Kings School,
but did not spill over it. The surface of this lake
crusted over with solid basalt as it cooled, but
before the whole lake solidified the underlying
liquid lava drained out, possibly as a result of it
breaching the tuff ring in the vicinity of Duke St
to the north. The surface crust of the lake
dropped 4–5 m as the lava drained out from
beneath. Where the lava lake cooled around its
edge, against the tuff ring and base of the scoria

◔ This large entrance to a lava cave within a Three Kings
lava flow lies in the backyard of a private house in
Landscape Rd.

cones, a bench of solid basalt was formed. This
bench was left behind as the lake drained and
can still be seen today beneath Three Kings
School and through many gardens along the east
side of St Andrews Rd between Mt Albert Rd
and Gorrie Ave. The beautiful stone walls in the
grounds of the school were constructed from the
rubbly surface of the old lava lake by unemployed
workers during the 1930s Depression.

After breaching the low crest of the tuff ring,
in the vicinity of present-day Duke St, lava poured
into a valley and flowed downstream for 3 km
before coming to a stop at Western Springs. The
valley filled with lava flows – their ridged and
rubbly surface can be followed today through
the back yards and streets in the suburbs of
Mt Eden South, Balmoral, Sandringham, St Lukes,
Morningside and Western Springs. These lava

❸ View from the southwest in 1949 across McCullough and Smallfield avenues to the partially quarried Three Kings scoria cones. South King (right) and East King (centre) are mostly removed while Big King (left) is protected within a reserve. *Whites Aviation Collection, Alexander Turnbull Library*
❹ The tuff layers in Liverpool St, Epsom, mantle an underlying sandstone ridge producing this upfold (anticline).

flows from Three Kings also contain some of Auckland's most accessible lava caves, although all have entrances on private property and require permission to gain access.

Human history

The pre-European Māori name for the cones is Te Tātua-o-Mataaho, or 'the war belt of Mataaho', and later adapted to become Te Tātua-a-Riukiuta (referring to the bringing together of the diverse genealogical lines of the local tribes by Riukiuta, another name of the leading tohunga, or priest, of the canoe *Tainui* when it visited Auckland about 600 years ago). The European name Three Kings derives from the Three Wise Men of the Nativity and was applied to the scoria cones by Captain Hobson's surveyor Felton Mathew in 1841.

Today, most of the Three Kings lava flows and tuff ring survive beneath myriads of suburban houses, shops and roads. The second-highest of the original scoria cones, named Big King, is the only one still standing, the remainder having been quarried away. Commercial quarrying at Three Kings began about 1913, but really took off after 1950 as Auckland's growth skyrocketed. By 2011, all the scoria resource within the quarry reserve had been extracted and at the present time the quarry is being partially filled and turned into a residential subdivision and playing fields.

In 1927, the Wesleyan Native School donated Big King and 5.5 hectares of land to the government as a reserve, thereby saving it from the diggers. Much of the rest of the land originally owned by the Wesleyan School inside the western portion of Three Kings explosion crater became a state-housing subdivision in the 1950s, centred around McCullough Ave.

Water supply
Since the eruption of Three Kings ceased, water has travelled underground through fractures and broken rock in its lava flows to the northwest before welling up at Western Springs to form the small lake in one of Auckland's most popular reserves. The first water supply for Auckland came from the Domain but by the 1870s there was an urgent need for another source of fresh water for the growing city and Western Springs was chosen. A brick pumphouse with a tall chimney was built to house large imported pumps that from 1877 were used to pump water to newly constructed reservoirs on the ridges of Karangahape Rd and Symonds St and after 1888 on Mt Eden. Western Springs water was supplied to all the inner suburbs until 1902 when it needed to be supplemented by water piped in from the Waitākere Ranges. In 1910, water from the newly constructed Waitākere Dam came on stream and the Western Springs system was retained for use as an emergency supply until 1956.

Three Kings Volcano also supplied water from inside its large explosion crater for local use. In 1915, a pumphouse was built near the base of South King to pump groundwater out of the porous scoria beneath for distribution around the district until 1923. This pumphouse is still present near the Three Kings shopping centre and is used by the Auckland Brass Band. Later, water was brought in by pipe from the Waitākere Ranges and stored in an underground reservoir installed in 1950 on the west side of Big King and a large concrete reservoir tank erected on the summit in 1953.

Liverpool Street tuff

Where it is: In the road cutting outside 27 Liverpool St, Epsom.

This is the best place to see volcanic ash erupted by wet explosive eruptions and base surges from Three Kings Volcano. Here the layers of tuff (hardened ash) form a fold over the crest of a buried sandstone ridge. Each layer was deposited from a separate eruptive blast or surge. Some layers contain numerous tiny volcanic hailstones that formed within the wet erupted cloud as it blasted out sideways (see page 13). Some layers are composed of more friable and easily eroded scoria that was erupted by dryer eruptive blasts. Towards the top of the tuff sequence there is a change to more massive beds of fine scoria that mark the switch to dry eruptions and the start of the scoria cone-building phase.

◐ Bedded tuff layers exposed in the Liverpool St road cutting have been cut by small faults, which appear to have displaced the layers as the tuff deposits compacted and moved slightly downhill on either side of the fold crest.

Puketāpapa/Pukewīwī/
Mt Roskill

● Puketāpapa/Mt Roskill
scoria cone with its artificial
flat top, the roof of a buried
water reservoir, viewed from
the south in 2018. *Photo by
Alastair Jamieson*

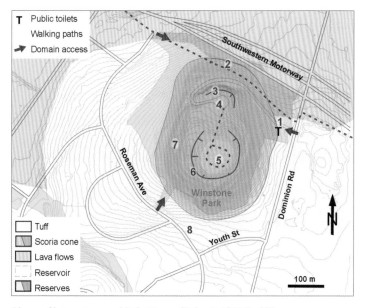

Places of interest around Puketāpapa/Pukewīwī/Mt Roskill:

1. Car park at main entrance to the
   reserve.
2. Section of the Southern Isthmus
   Cycleway and walkway alongside
   Southwestern Motorway.
3. Road cuttings through loosely
   consolidated scoria.
4. Car park over flat infilled northern
   crater.
5. Flat top of buried concrete water
   reservoir fills the enlarged southern
   crater of the volcano.

6. Viewing point on circular rim of
   southern scoria cone. The crest
   retains terraces and kūmara pits
   from pre-European Māori use.
7. Extensive pre-European terracing
   around upper western and
   southern slopes of the cone.
8. Gentler slopes around southwest
   side of scoria cone (Roseman Ave)
   are the unburied remnant of the
   volcano's tuff ring.

Land status: Almost all of the cone is publicly accessible in Winstone Park, administered by the Tūpuna Maunga Authority.

What to do: Walk around the summit for views of the surrounding suburbs and the earthwork remains of the large, pre-European Māori pā on the volcano.

Geology

Puketāpapa/Mt Roskill has been dated using the argon-argon method at approximately 105,000 years old. Initial wet explosive eruptions built a tuff ring, part of which can still be seen as a low arcuate ridge around the southwest side of Mt Roskill. The rest of the tuff ring was buried by the later cone. Dry fountaining eruptions from two vents built the 55-metre-high oval scoria cone (109 m ASL). The two original shallow craters are no longer visible because of the reservoir installation and parking area

◐ Aerial view north across the top of Puketāpapa/Mt Roskill in 1958 showing its twin craters, double cone and extensive terracing and pits, prior to its extensive damage when the water reservoir was installed. *Whites Aviation, University of Auckland*

❷ In 1961–62, a large circular concrete reservoir was buried in the top of Mt Roskill; 25,000 cubic metres of scoria was excavated to create the hole for the reservoir, which was then buried. *Sir George Grey Special Collections, Auckland City Libraries*

now at the north end of the summit. Lava exuded from the northern and western base of the scoria cone, flowing 500 m north, almost as far as Mt Albert Rd, and 2 km west down the upper reaches of Oakley Creek valley almost as far as Hendon Ave, Ōwairaka. This lava flow, and those further downstream from Mt Albert, partially dammed the upper Oakley Creek, creating the swampy areas after which Ōwairaka ('the water of Raka') was named.

Human history

This volcano has several Māori names – Pukewīwī, meaning 'hill covered in rushes', and Puketāpapa, meaning 'flat-topped hill'. This was another of the pre-European cone pā of Auckland, with extensive terracing and storage pits, although many were lost when the water reservoir was installed in the craters. The European name is believed to have been given by the earliest European owner, Alexander Kennedy, after Roskhill in Scotland close to his original home on the Isle of Skye. In the 19th century it was also referred to as Mt Kennedy.

European owners of Mt Roskill scoria cone have been Alexander Kennedy (1844–49), Joseph May (1849–91) after whom May Rd is named,

and George Winstone (1891–1928). In 1928, George Winstone gave Mt Roskill (9 hectares) to the Mt Roskill Borough Council to be a recreation reserve, which was named Winstone Park Domain. The road access to the top of the volcano was completed when the water reservoir was buried in the southern crater in 1961–62. Tennis courts and a croquet lawn were developed within the park at the cone's northern base but these were sacrificed when the Southwestern Motorway was constructed through them.

In the 2000s, Transit New Zealand proposed a substantial vertical cut through the lower northern slopes of the cone as part of the motorway extension. The Auckland Volcanic Cones Society challenged this proposal in the Environment Court and, as a compromise, the road was constructed further away and the new cutting was contoured to appear to be a continuation of the cone's slopes.

Te Ahi-kā-a-Rakataura/ Ōwairaka/Mt Albert

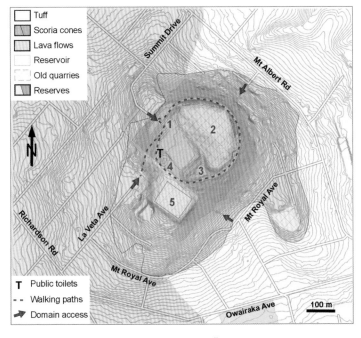

Places of interest on Te Ahi-kā-a-Ōwairaka/Mt Albert:

1. Main entrance and parking area.
   Two disused reservoirs are buried
   on either side of the gateway.
2. Site of highest part of original cone
   quarried by NZ Railways for ballast
   using an incline tramway. Flat floor
   of disused quarry is now used by
   Mountain Green Archery Club.
3. Panoramic lookout from the
   trig, now the highest remaining
   part of the cone (133 m ASL).
   Pre-European terracing and pits
   around the summit and down
   the eastern slopes.
4. Parking area beside soccer field on
   flat floor of another major quarry.
   In the trees beside the car park are
   several pre-European kūmara pits.
5. Flat grassed area on top of
   buried water reservoir that was
   constructed inside the excavated
   crater of the secondary southern
   scoria cone.

◔ The top of Mt Albert scoria cone has been lowered
and extensively modified by quarrying (centre and right)
and the installation of a flat-topped water reservoir (left).
Photo by Alastair Jamieson, 2009

⬆ John Guise Mitford's 1845 watercolour of Mt Albert from the north, before quarrying had removed its top. *Alexander Turnbull Library*

Land status: The top half of the scoria cone is in Mt Albert–Ōwairaka Domain, administered by the Tūpuna Maunga Authority. The lower slopes and lava-flow fields are mostly covered in privately owned dwellings.

What to do: Walk around the sealed circular roadway to the summit trig and enjoy the views.

Geology

Mt Albert is the westernmost volcano in the Auckland Volcanic Field and has been dated at about 120,000 years old. It is the remains of a large scoria cone (formerly with a conical top 148 m high) that has had the top 15 m removed by quarrying. The earliest eruptions were of the wet explosive style and produced a tuff ring, most of which has been buried by the large, complex scoria cone that was subsequently thrown up during dry fountaining eruptions. The eastern remnants of the tuff ring form the ground surface beneath parts of Mt Albert Rd and the Mt Albert Research Centre. The scoria cone had a large breached crater opening to the northwest and a smaller crater on the mound in the southwest.

Lava extruded out from around the lower flanks of the cone. Some flowed south, blocking the Oakley Creek valley and forming the Ōwairaka swamp that now underlies the area around Stoddard Rd. More lava flowed north down the Oakley Creek valley with its toe reaching to where the present-day shore of the Waitematā Harbour is located in the vicinity of the Waterview Motorway Interchange. Vegetation and debris from the Ōwairaka swamp soon blocked the cracks in the lava flow and the swamp overflowed down a new course along the western edge of Mt Albert's flows, forming the present-day Oakley Creek. Lava also flowed north from the breached crater around the end of present-day Summit Drive and into the Meola Valley near Chamberlain Park Golf Course. A number of lava caves are known within these flows from Mt Albert.

Human history

The best-known Māori name for this volcano is Ōwairaka ('the dwelling place of Wairaka'). Tainui sources, however, apply the names Te Wai inu roa-o-Raka ('the drinking waters of Rakataura') and Te Ahi ka roa-a-Raka ('the long burning fire of Rakataura') to the mountain and the surrounding area. Rakataura (or Riukiuta) was the leading tohunga on the canoe *Tainui* when it visited Auckland about 600 years ago. The scoria cone was extensively modified by pre-European Māori with the construction of terraces, pits, ditches and banks as part of a

◑ North side of Mt Albert in the 1890s. To help build the Kaipara railway in the 1880s, New Zealand Railways built a branch line to the base of the cone in Toroa St. A self-acting incline tramway was constructed down these northern slopes from the base of the quarry in the picture. Two trucks operated at a time on a single track. The weight of the full scoria truck running down the slope pulled the empty truck up. They passed each other at a twin-tracked passing bay in the middle. *Auckland Museum*

defensive pā. Today, only a few remnants of
these survive. Prominent Ngāti Whātua chief
Apihai Te Kawau accompanied missionary
Samuel Marsden to the summit in 1820.

Mt Albert was named after England's reigning
Queen Victoria's consort Prince Albert in early
colonial days. In the 19th century, small springs
emanating from beneath a Mt Albert lava flow
near the tidewaters in Oakley Creek were used to
supply water for nearby Avondale and for a while
to supplement the Auckland City supply.

Mt Albert's scoria cone has had a long history
of quarrying. Around 1.5 million cubic metres of
scoria was removed between 1860 and 1959 to
provide material for roads, railway-line ballast
and the Northwestern Motorway construction.
The New Zealand Railways scoria ballast
pit on the northern part of the summit was

opened in the 1880s and closed in 1928, but
quarrying continued in the summit area (now a
soccer field). The floors of these quarries have
been rehabilitated as grassed playing fields
around the top of the mountain. Another small
quarry on the southwest corner of the mountain
(end of La Veta Ave) was active in the 1950s.

Mt Albert is a landmark for the district, and
decades of persistent public protest eventually
succeeded in stopping the quarrying. In 1903,
central government set aside 4.9 hectares for
a domain and in 1931 the Railways Department
added another 3.6 hectares. In 1946–52, there
was more damage when a large water reservoir
was buried on the summit of the secondary
southwestern cone, giving Mt Albert its
unnaturally flat grassed top.

● View from One Tree Hill showing the pre-quarrying eastern profile of Mt Albert in 1905. *Courtesy of Auckland Libraries Heritage Collections*
● Road cutting in Unitec grounds through columnar-jointed basalt of a Mt Albert lava flow, 2007.

● View north across the decapitated top of Mt Albert in 1958. The quarry in the foreground (now the end of La Veta Ave) was the last to operate. The flat area in the right foreground is the top of a buried circular reservoir whereas the two flat areas beyond are former quarries. *Whites Aviation Collection, Alexander Turnbull Library*

Te Hopua-a-Rangi/
Gloucester Park

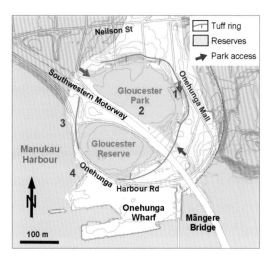

Places of interest at Te Hopua:

1. Main entrance and parking area for Gloucester Park.
2. Infilled floor of Te Hopua crater, formerly a tidal basin.
3. Aotea Sea Scouts building with intertidal exposure of blocky breccia that resulted from initial phreatic (steam) eruptions from Te Hopua which blasted through One Tree Hill lava flows.
4. Original breach in tuff ring where the tide flowed in and out of the crater basin.

◔ View south over Te Hopua explosion crater on the Onehunga foreshore, now bisected by the Southwestern Motorway to and from the airport, with Gloucester Park sports fields on the northern (near) part of the reclaimed crater floor. *Photo by Alastair Jamieson, 2018*

⊘ Sloping intertidal platform of bedded brown tuff that formed the circular tuff ring around Te Hopua explosion crater. Here the tuff contains large, angular blocks of grey basalt which were part of a lava flow from One Tree Hill that was blasted apart and thrown into the air by the wet explosive eruptions. Behind is the historic boating clubhouse now used by the Aotea Sea Scouts.

Land status: The crater floor is publicly accessible Gloucester Park, except for the motorway that bisects it. The tuff ring is either public road or privately owned properties.

What to do: Walk around the inside of the northern part of the crater; walk around the crest of the tuff ring on the footpaths from Neilson St to the Aotea Sea Scouts clubhouse or walk on the grass around the wetlands south of the motorway.

Geology

The 500-metre-diameter Te Hopua explosion crater and surrounding low tuff ring is one of Auckland's smaller volcanoes. Coring in the middle of the crater has shown that it erupted before 30,000 years ago, blasting its way through a cap of One Tree Hill lava flows. Following the eruptions the crater became a freshwater lake and sediment slowly accumulated on its floor. About 8000 years ago, as sea level rose after the end of the Last Ice Age, the low tuff ring on the south side was breached by the tides and the lake became a tidal basin on the Manukau Harbour coast. The basin rapidly filled with marine mud and was largely intertidal when humans arrived. Much of the low tuff ring around the crater's north and east sides is still present although hidden beneath a number of industrial buildings and a motel.

Human history

Te Hopua-a-Rangi, meaning 'the basin of Rangi', is thought to refer to the Waiohua ancestress Rangihuamoa. Early European settlers called this Geddes Basin after one of the earlier colonial owners, Alexander Geddes. Onehunga's first wharf was built off the end of the western arm of the tuff ring in 1858 and the basin was used as the Onehunga boat harbour with sheltered moorings for shallow-draught vessels. In 1878, a new wharf with rail links to Onehunga Station was opened at the end of the eastern arm, where Onehunga Wharf is today. In 1911, the two-storey wooden clubhouse for the Manukau Yacht and Motor Boat Club was erected on the outer edge of the western arm.

In the early 1930s, the breached entrance into the basin had silted up with mud and over

● Plan of properties around Te Hopua crater when it was still a tidal boat harbour, 1862. Note the location of the first wharf. Most of the property boundaries on the north and east sides are still current or have been subdivided.

the next decade (up to 1941) it became the Onehunga Council tip, which was later levelled and grassed. The new sportsground was named Gloucester Park, after His Royal Highness the Duke of Gloucester, who visited New Zealand in 1934. In the late 1930s–40s, some of the reclaimed land was used for the Onehunga Speedway and, during the Second World War, US Army barracks were constructed on the park.

In the 1970s, the Southwestern Motorway (State Highway 20) to Auckland Airport was built through the centre of the reclaimed Te Hopua explosion crater. It was widened further in 2009.

● Te Hopua explosion crater was used as the Onehunga boat harbour, as seen here in the 1920s, until it was reclaimed in the 1930s. In the top left is the two-lane Māngere Bridge (1914–83) across the Manukau Harbour. It was built to replace the first single-lane, wooden bridge (1875–1914). *Breckon Album, Auckland Museum*
● By 1954, after reclamation was complete, part of the floor of Te Hopua explosion crater was used as a timber yard. Part of the earlier Onehunga Speedway track is also visible. *Whites Aviation, University of Auckland*

Rarotonga/Mt Smart

⊙ Rarotonga/Mt Smart has been removed by quarrying and the site converted into Mt Smart Stadium with the remnants of the scoria cone's lower slopes around the south and eastern sides planted in pōhutukawa trees. *Photo by Alastair Jamieson*

⊙ Thomas Hutton's watercolour of the Manukau Harbour from the summit of Mt Wellington, 1870s. Mt Smart is the cone on the left with Puketūtū Island behind. Te Hopua crater and tuff ring form the small peninsula near the centre. *Auckland Public Art Gallery*

Land status: The site of the quarried-out cone is all within Mt Smart Stadium, which is managed by Auckland Council on behalf of the Tūpuna Maunga Authority. Most of the lava-flow field is privately owned industrial property except the coastal fringe.

What to do: There is not much to do unless you come for a sporting or cultural event at the stadium. Walk around the perimeter of the former cone along Maurice Rd, Mt Smart Rd and Beasley Ave and back through the stadium car parks if the gates are open. Some of the toes of the lava flows can be seen amongst the mangroves alongside the coastal walkway.

Geology

Rarotonga/Mt Smart erupted through the southeast edge of the One Tree Hill lava-flow field almost exactly 20,000 years ago. Initial wet explosive eruptions were followed by major fountaining that built an 87-metre-high scoria cone with a small central summit crater. Lava flowed out from the base of the cone and spread east and south, forming a small lava-flow field covering 300 hectares. The toe of some of the flows now forms the foreshore of the Māngere arm of the Manukau Harbour.

◔ Rarotonga/Mt Smart scoria cone prior to any significant quarrying. *Hugh Boscawen, 1899, Auckland Museum*

◑ By 1949, quarrying had removed much of Mt Smart scoria cone. A branch rail line from the Onehunga line extended right into the quarry. Here rail wagons are lined up waiting to be filled with scoria from the crusher. *Whites Aviation, University of Auckland*

◒ This pahoehoe ropey lava surface in the mangroves adjacent to the Onehunga–Southdown coastal walkway is developed on the toe of a lava flow from Mt Smart.

Human history

This volcano has at least two Māori names – Te Ipu kura-a–Maki, meaning 'the red bowl of Maki, a warrior' (in reference to the summit crater), and Rarotonga, meaning 'the lower south' (a name brought from the ancestral homeland, Hawaiki, and placed on the mountain as a reminder). Prior to the arrival of Europeans, the scoria cone was extensively terraced and used as a defensive pā by local Māori.

Mt Smart was given its European name in the 1840s by early colonial settlers after Henry Dalton Smart, a lieutenant in the New Zealand Mounted Police. The earliest quarrying began in 1865 to obtain scoria ballast for some of Auckland's first railway lines. From then until the 1960s, New Zealand Railways excavated the scoria and transported it away in rail wagons that were hauled right into the quarry along a branch railway line. In the 1940s, pōhutukawa were planted on the remnant lower slopes on the southern and eastern sides of the cone and

today these hint at the original size of Mt Smart's substantial cone. A sports stadium was opened in the floor of the old quarry in 1967 and this was further developed as the main stadium for the 1990 Commonwealth Games. In the early 21st century, Mt Smart Stadium has been home to the New Zealand Warriors rugby league team and the Big Day Out music festival, as well as many other large music and cultural events.

Ōrākei Basin

◐ Ōrākei Basin explosion crater and surrounding tuff ring from the south in 2018. The grassy Kepa Rd slopes on the north side of the crater are draped in tuff that is unstable and slowly sliding back down into the crater and Purewa Creek. *Photo by Alastair Jamieson*

Land status: The basin has a narrow coastal reserve with a walkway around the inner shore, managed by Auckland Council. Most of the higher parts of the tuff ring are in private properties.

What to do: Launch your own yacht, canoe or boat from the boat ramp and explore the basin and its Waiatarua arm from the water. The most popular activity is to walk the track around the basin, which includes two steep climbs (about 40 minutes).

Geology

Ōrākei Basin is a large, 800-metre-diameter explosion crater that erupted on the side of Purewa Creek. Ejected ash and base surges built up a tuff ring around three sides but on the northeastern side the ash was plastered onto the steep northern slopes of Purewa Valley. In wet weather these tuff deposits are prone to slumping and thus the unstable Kepa Rd slopes are undeveloped and kept in grazed pasture.

In 2016, a 100-metre-long core was obtained from the basin sediment. It shows that the

crater was formed by eruptions approximately 120,000 years ago. Sea level was as high as today (Last Interglacial period) but the crater was not breached by sea water at that time. Core sampling also indicates that after eruption the crater became a lake and over the next 110,000 or so years, it slowly filled with sediment, shallowing to become a freshwater swamp. Rising sea level allowed the sea to flood the swamp 9000 years ago and transform it into a tidal basin that filled with sandy marine mud, and by the time of human settlement it was intertidal sand flats.

Human history

Ōrākei literally means 'the dwelling place of Rakei-iri-ora', who was a descendant of Rakataura (or Riukiuta), the leading tohunga on the canoe *Tainui*. Pre-European Māori lived around the shore of the basin and built a defended pā on the steep-sided crest of the

◉ View north across Ōrākei Basin from Upland Rd in 1921 prior to construction of the railway embankment through it. The southern edge (left foreground) was fringed with low mangroves and salt marsh. In the early 1930s, 20 Depression relief workers spent a year clearing the mangroves and constructing a grassy picnic area and paths. *James D. Richardson, Sir George Grey Special Collections, Auckland City Libraries*

◉ View south across Ōrākei Basin explosion crater, 1921, showing the area now occupied by a car park and boat ramp. The crest of partially quarried Maungarahiri/ Little Rangitoto can be seen on the left skyline. *James D. Richardson, Sir George Grey Special Collections, Auckland City Libraries*

◉ Vast quantities of shell midden can be seen beside the road to the car park inside Ōrākei Basin – a reminder of the extensive use pre-European Māori made of this beautiful location.

northwestern arm of the tuff ring (Orakei Rd peninsula).

Following the arrival of Europeans, the suburbs of Auckland gradually spread around Ōrākei Basin and in the mid-1920s the main trunk railway line was constructed through the northern portion of the crater. A wide trench was dug through the Orakei Rd portion of the tuff ring and the debris used to help construct the railway embankments across the tidal flats. The embankment created a shallow, artificial saltwater lake with control gates under the railway line. Today these are opened twice a month during spring high tides to flush out the nutrient-rich brackish water and replace it with sea water. In 1933, the *NZ Herald* proclaimed that Ōrākei Basin was the 'largest controllable swimming pool in New Zealand with public bathing sheds at the bottom of Lucerne Rd'.

In 1929, a tunnel was dug under Remuera Rd ridge to drain Lake Waiatarua into Ōrākei Basin. In the late 20th century, this tunnel carried much suspended clay pumped out from the Mt Wellington quarry (now Stonefields subdivision), which settled in the basin and was blamed for promoting algal blooms that rotted, producing foul smells. In the early 2000s, the accumulated mud on the floor of Ōrākei Basin was dredged and removed. Today, Ōrākei Basin is home to the Auckland Water Ski Club, Orakei Sea Scouts and Akarana Young Mariners.

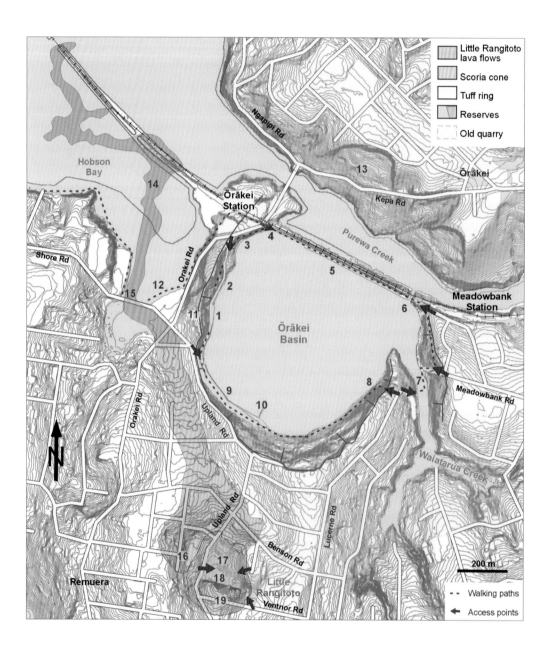

◔ *Places of interest around Ōrākei Basin and Maungarahiri/Little Rangitoto:*

1. Car park and boat ramp inside Ōrākei Basin explosion crater.
2. Orakei Sea Scouts and Akarana Young Mariners clubhouses. Note shell midden in road cuttings nearby.
3. Walkway through bush alongside basin.
4. Walkway from Ōrākei Station; exposures of bedded tuff under railway bridge.
5. Shared wooden walkway and cycleway alongside railway embankment.
6. Cutting through ridge for basin control gates. Note bedded tuff in cutting.
7. Walkway through bush and across walking bridge over Waiatarua arm; some steep sections with steps.
8. Auckland Water Ski clubhouse with steep steps to Lucerne Rd behind.
9. Grassed picnic area reclaimed from mangroves in 1930s.
10. Low stone jetty built in 1932 for use by Ōrākei Basin Model Yacht Club members.
11. Views from grassy path along western crest of tuff ring; earthwork remains from pā and European buildings.
12. Boardwalk through mangroves as part of walkway around fringes of Hobson Bay.
13. Kepa Rd passes across steep grassed slopes underlain by unstable Ōrākei Basin tuff.
14. Little Rangitoto lava flow beneath marine sediment in Hobson Bay.
15. Surface of lava flow amongst low mangroves on seaward side of Shore Rd next to stream bridge.
16. Solid basalt of Little Rangitoto lava flow in wall and steep front garden of 10 and 16 Raumati Rd.
17. Little Rangitoto Reserve. Partly infilled floor of quarried scoria cone. Children's playground, flying fox and concrete skateboard bowl.
18. Old quarry face of scoria from southern slopes of Little Rangitoto cone. See fused scoria and large smooth volcanic bomb.
19. Ventnor Rd goes up and over lower slopes of former scoria cone.

◔ It is possible to walk right around Ōrākei Basin's drowned crater on a network of paths, steps and recently constructed boardwalks. This bridge crosses the drowned mouth of Waiatarua Stream where it flows into the basin.

Maungarahiri/
Little Rangitoto

⊕ By 1899, Little Rangitoto scoria cone was already severely damaged by quarrying. This view is from the Ōrākei Basin tuff ring crest along Upland Rd past the Benson Rd corner to Little Rangitoto. *Hugh Boscawen, Auckland Museum*
⊙ The same view in 2019 as in Boscawen's 1899 photo along Upland Rd to the remains of Little Rangitoto Volcano.
⊖ Little Rangitoto Reserve in Upland Rd, Remuera, in 2009, is the partially filled quarry that removed most of Maungarahiri/Little Rangitoto's scoria cone. Remnants of the cone's southern slopes underlie the houses on the right.
Photo by Alastair Jamieson

Land status: Parts of the cone site are in Little Rangitoto and Ventnor reserves, managed by Auckland Council. The remainder of the cone and upper half of the lava flow are in private houses.

What to do: Bring the kids to the playground or skateboard bowl, or to learn to ride their bikes on the circular path. Walk around the former volcano, from the playground down the path to Crown Lane and up through Ventnor Reserve and Ventnor Rd and back along Upland Rd. Walk out on the lava-flow surface beside Shore Rd, but wear gumboots.

Geology

Maungarahiri/Little Rangitoto Volcano in Remuera was a relatively small, 70-metre-high scoria cone that erupted about 24,500 years ago, tens of thousands of years after neighbouring Ōrākei Basin. A lava flow oozed from the north-western base of the scoria cone and flowed as a narrow ribbon of fluid lava down a small valley around the west side of Ōrākei Basin tuff ring and northwards into the lower Purewa Valley. At the time, sea level was lower than now and today's Hobson Bay was forested. Today, the northern end of Little Rangitoto lava flow is buried beneath the mud of Hobson Bay. Its sinuous route out beneath the railway embankment has been traced by remote magnetic sensing techniques by geophysicist Matt Watson (see map on page 174).

⊙ Thin layers of vesicular basaltic lava within scoria exposed in former quarry face in Little Rangitoto Reserve. These are rootless flows because the molten lava fountained out of the vent but was still liquid when it hit the scoria cone surface and began to flow off downslope.

○ View north over Little Rangitoto's scoria quarry in 1931. Upland Rd is on the left. *James D. Richardson, Sir George Grey Collections, Auckland Libraries*

Human history

The Māori name is Maungarahiri, meaning 'hill of the sun's rays'. Presumably, Little Rangitoto was given this name by early Europeans because of the superficial resemblance of its profile to its big cousin across the water. Most of the scoria cone was quarried away between the late 19th century and the 1950s. The hole was subsequently used for refuse disposal and then rehabilitated as the grassed Little Rangitoto Reserve. Houses that overlook the reserve on the north side of Ventnor Rd are built on what remains of the southern slopes of the scoria cone.

Places of interest around Maungarahiri/ Little Rangitoto:
See pages 174–75.

○ The surface of the Little Rangitoto lava flow can be seen among the low mangroves on the edge of Hobson Bay beside Shore Rd bridge.

⊙ Maungarei/Mt Wellington scoria cone with Panmure Basin's breached explosion crater beyond. *Photo by Alastair Jamieson, 2018*

Volcanoes of eastern Auckland

Fifteen volcanoes are located on either side of the Tāmaki Estuary and extend from St Heliers in the north to Ōtāhuhu in the south. All but two show evidence of initial explosion craters and surrounding tuff rings. Twelve produced scoria cones, ranging in size from the small cone in Hampton Park to the tallest cone on the isthmus, Maungarei/Mt Wellington. Four have completely lost their cones to the ravages of quarrying – Purchas Hill, McLennan Hills, Green Mount and Ōtara Hill. Mt Wellington, McLennan Hills and Green Mount produced extensive lava-flow fields, some of which have been quarried for aggregate, but large areas still underlie residential and industrial subdivisions. The eruption of McLennan Hills and Mt Richmond 30,000–40,000 years ago, at a time when sea level was lower, is believed to have dammed the former Tāmaki River and diverted its flow from westwards into the Manukau Valley to northwards into the Waitematā Valley, its present course (see map on page 24). The remains of all but three of these volcanoes are in public reserves.

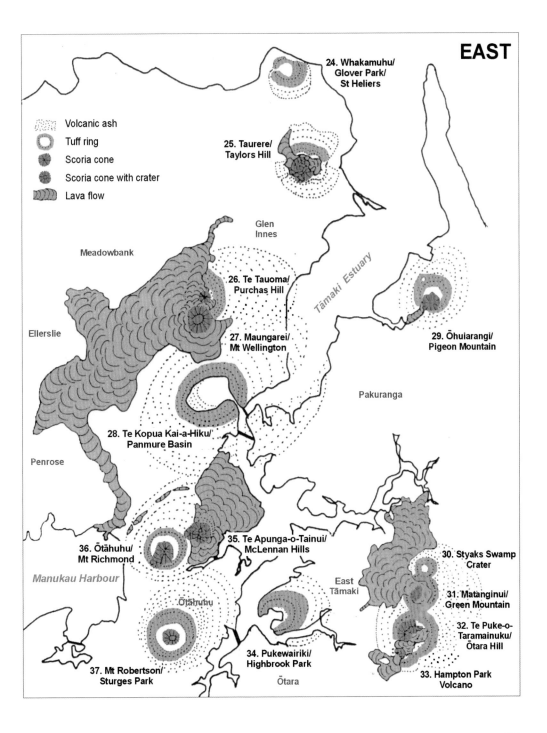

EAST

Volcanic ash
Tuff ring
Scoria cone
Scoria cone with crater
Lava flow

24. Whakamuhu/
Glover Park/
St Heliers

25. Taurere/
Taylors Hill

Glen
Innes

Meadowbank

26. Te Tauoma/
Purchas Hill

Tāmaki Estuary

Ellerslie

27. Maungarei/
Mt Wellington

29. Ōhuiarangi/
Pigeon Mountain

Pakuranga

28. Te Kopua Kai-a-Hiku/
Panmure Basin

Penrose

35. Te Apunga-o-Tainui/
McLennan Hills

36. Ōtāhuhu/
Mt Richmond

30. Styaks Swamp
Crater

Manukau Harbour

East
Tāmaki

31. Matanginui/
Green Mountain

Ōtāhuhu

32. Te Puke-o-
Taramainuku/
Ōtara Hill

34. Pukewairiki/
Highbrook Park

37. Mt Robertson/
Sturges Park

Ōtara

33. Hampton Park
Volcano

181

Whakamuhu/
Glover Park/St Heliers

◔ View from the north over Whakamuhu/
Glover Park explosion crater (in bowl-shaped
Glover Park) and surrounding house-covered
tuff ring, 2009. Water tower is on far (southern)
crest of the tuff ring. St Heliers Beach is in top
right. *Photo by Alastair Jamieson*

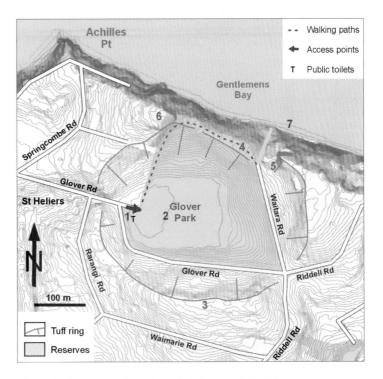

Places of interest around Whakamuhu/Glover Park/St Heliers:

1. Main Glover Park car park, children's playground, site of stream overflow from crater swamp.
2. Playing fields on drained swampy floor of explosion crater.
3. Water reservoir on concrete tower on highest part of southern crest of tuff ring.
4. Path along northern crest of tuff ring with panoramic views of Motukorea and Rangitoto volcanoes.
5. High cutting through tuff ring in abandoned house site at 27 Waitara Rd. View through fence.
6. Site of pre-European pā under houses on high point of tuff ring.
7. Blocks of tuff on beach; accessible from Karaka Bay at low tide.

◐ View from the southwest across Glover Park explosion crater in 1958 after the sports field had been formed and residences built around the tuff ring crest. *Whites Aviation, University of Auckland*

◑ View west across Whakamuhu/Glover Park explosion crater in 1899 with the high crest of the tuff ring forming the hill beyond. Coastal erosion on the north side (right) has removed some of the tuff ring. *Hugh Boscawen, Auckland Museum*

Land status: The floor and some of the inner slopes of the crater are in Glover Park, administered by Auckland Council. Most of the rest of the slopes and tuff ring crest are in private houses and roads.

What to do: Play ball games on the sportsground when they are not in formal use; children's playground by the car park; take a short walk around the northern part of the tuff ring with views over the harbour, Rangitoto and Motukorea.

◓ Small block of schistose basement rock in tuff from the sea cliffs below Glover Park.

Geology

Glover Park in St Heliers is located inside a 500-metre-diameter explosion crater. The surrounding tuff ring is being eroded by the sea on the north side, forming high, pōhutukawa-clad sea cliffs. A water reservoir on a concrete tower is located on the crest of the tuff ring amongst houses on the south side, and an overflow stream has eroded an outlet valley on the northwestern corner (Glover Rd). Like other explosion craters, this one would have originally been a freshwater lake, progressively filling with sediment to become a swamp. Recent geochemical finger-printing studies on basaltic ash layers in a core through this sediment indicate that Glover Park

Volcano erupted between 140,000 and 180,000 years ago – making it one of the oldest volcanoes in the Auckland Volcanic Field.

Blocks of bedded tuff lie at the foot of the eroding sea cliffs between Achilles Pt and Karaka Bay. Some contain angular lumps of a variety of hard, green and grey schist, greywacke and gabbro rocks that were thrown out of the volcanic vent during eruptions and were presumably ripped from the walls of the volcano's neck at considerable depth by explosive blasts. These rocks are unlike any others seen at the surface around Auckland and are samples of the hard Paleozoic basement rocks under this area. They come from close to the eastern margin of a belt of highly magnetic basement rocks (Dun Mt Ophiolite Belt) that has been mapped using aeromagnetic methods by the University of Auckland's Jennifer Eccles and shown to underlie Auckland in a number of northwest-oriented belts.

Human history

The Māori name of this volcano is Whakamuhu, meaning 'to lead into a thicket', referring to its swampy crater floor. The European names include Glover Park, named after Albert Edward Glover, city councillor and Member of Parliament in the 1890s–1910s. It is also commonly called St Heliers Volcano after the local suburb, which derives its name from St Helier, the capital of Jersey, in the Channel Islands, United Kingdom. The crater was acquired for a park in the 1930s and in the 1950s the swampy floor was drained and filled. The crater floor is now used for sports, such as cricket, athletics and soccer. In 1953, the Tāmaki Ex-Servicemen's Women's Auxiliary planted trees in the park to commemorate the men of the district who lost their lives during the two world wars.

Taurere/Taylors Hill

◑ View from the northwest across Taurere/Taylors Hill scoria cone complex in 2009. *Photo by Alastair Jamieson*

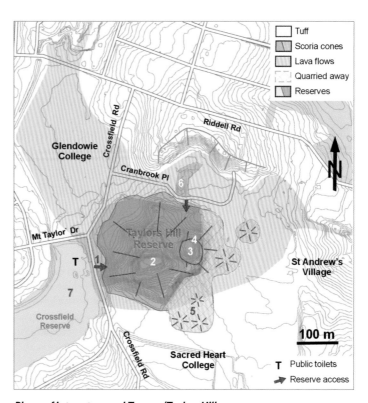

Places of interest around Taurere/Taylors Hill:

1. Main entrance to Taylors Hill Reserve.
2. Summit of scoria cone and Taurere pā.
3. Main fountaining bowl-shaped crater.
4. Grove of karaka trees possibly descended from trees in Māori oral history.
5. Site of quarried-away scoria mounds in playing fields of Sacred Heart College.
6. Small reserve at Cranbrook Place on lava flow between scoria cone and tuff ring.
7. Crossfield Reserve on drained swamp that was dammed by volcano's eruption.

Land status: The central cone area is in Taylors Hill Reserve, managed by Auckland Council. Most of the tuff ring, explosion crater floor and lava flows lie beneath suburban homes and the playing fields of Sacred Heart and Glendowie colleges.

What to do: Walk to the top of the cone to admire the view and imagine the extent of the tuff ring, crater and lava flows.

Geology

Taurere/Taylors Hill in Glendowie erupted in the middle of a valley that drained east to the Tāmaki River. The first eruptions were wet explosive style, throwing out pulverised rock and volcanic ash and creating a 900-metre-diameter crater partly surrounded by a tuff ring. No tuff ring built up on the south side as the ash mantled the existing slopes of a sandstone ridge, now the site of Sacred Heart College. The low tuff ring rim to the east and west is either buried or breached by later flows, with the thickest tuff forming the prominent hill on the crater's north side.

A complex of cones and mounds was built from scoria erupted from at least five separate vents within the crater. The highest point (56 m ASL) is the trig on the remaining part of the hill. Lava oozed from the eastern and western base of the main scoria cone complex. To the east, a small flow rafted away some of the scoria, with a narrow tongue escaping the explosion crater and flowing down a narrow valley beneath present-day St Andrew's Village towards Glendowie Bay. To the west, lava flowed over the tuff ring rim and then north under present-day Glendowie College before coming to a stop near Riddell Rd. The eruptions at Taylors Hill blocked the east-flowing valley, creating a swampy wetland, now drained and used as Crossfield Reserve. The stream was diverted to the north around the tuff ring. Paleomagnetic studies indicate that this was one of a group of five Auckland volcanoes that erupted during the magnetic excursion about 30,000 years ago.

❹ Taurere/Taylors Hill scoria cone from Mt Taylor Drive in the west, 2010.

⊙ View from the south across
Taurere/Taylors Hill Volcano in 1958,
prior to suburban subdivision or the
construction of adjacent Glendowie
College. Quarrying was removing
scoria mounds to the south and east
of the main cone – now the site of
Sacred Heart College playing fields.
Whites Aviation, University of Auckland

Human history

The Māori name is Taurere, meaning 'loved
one flown away'. Remaining parts of the scoria
cone have numerous pre-European terraces and
pits, part of an extensive Māori pā site. Taylors
Hill was given its European name in the 1850s
after the then landowner Richard James Taylor
and his Glen Taylor Farm. Taylors Hill Reserve,
encompassing the main scoria cone area, was
donated to the city by the estate of William
Innes Taylor (Richard James' brother) in 1924.
Quarrying began on Taylors Hill scoria cones in
the early 20th century but most of the damage
occurred in the 1950s and 1960s when the cones
and mounds around the south and east sides
were removed.

Te Tauoma/Purchas Hill

◔ View south over the site of Purchas Hill explosion crater (foreground) with Maungarei/Mt Wellington cone built over the southern part of the tuff ring. The site of the quarried-away Purchas Hill scoria cones in the middle of the crater is at the far end of the undeveloped land in the centre. *Photo by Alastair Jamieson, 2018*

Land status: The site of the northern scoria cone is shown on the Unitary Plan as intended open space reserve whereas the southern cone's footprint lies under College Rd and private land. The site of the tuff ring and crater is almost all in industrial subdivision and road except for linear Morrin Reserve, which approximates the crest of the northern part of the tuff ring.

Geology

Te Tauoma/Purchas Hill volcanism began with a series of wet explosive eruptions that blew out a shallow, 600-metre-wide crater and built up a low surrounding tuff ring of expelled rock and ash. The tuff ring is still preserved on the northern and eastern sides with its crest approximately along the routes of Elizabeth Knox Place and Tainui Rd. Its southern sector was later buried by eruptions from Mt Wellington.

Activity later switched to dry fountaining of frothy lava from two adjacent vents, building up a small, 30-metre-high, double scoria cone in the centre of the explosion crater. Fine ash within the upper parts of the scoria cone is thought to be derived from the earliest wet explosive eruptions from an awakening Mt Wellington, 500 m to the south. Some of Mt Wellington's voluminous lava flows poured into the western side of the Purchas Hill explosion crater.

In the past, geologists lumped Purchas Hill in with Mt Wellington and counted them as one volcano. More recently they have been treated separately but they are probably one of

the volcano pairs in Auckland that erupted one after the other. There appears to have been no significant time gap between the two volcanoes' eruptions. It is likely that the Mt Wellington magma came up the same conduit as that used by the Purchas Hill magma. The age of both volcanoes is close to 10,000 years.

Human history

The Māori name Te Tauoma means 'to arrive running'. Like all other cones in Auckland, Te Tauoma's double cone was terraced and fortified as a small defended pā by pre-European Māori.

Ferdinand von Hochstetter gave the cone its European name in 1859, after the Rev. Arthur Guyon Purchas, who accompanied him on a number of his excursions around Auckland and its volcanoes. Quarrying began as early as the 1850s to supply scoria for the main road to Panmure, which passed right through the explosion crater on its way from Auckland via St John's College at the end of Remuera Rd ridge. Quarrying continued until scoria ran out in 1966 but it remained as a council works depot for many years after that. Part of the tuff ring in the east was quarried to supply clay to an on-site brickworks in the 1930s. Most of the remainder of the tuff ring was flattened during preparation for the industrial subdivision on the Knox estate in the 1970s. As part of the Stonefields housing

⊙ View north from Mt Wellington over the northern arc of Purchas Hill's tuff ring (foreground) in 1930. Morrin Rd cuts through the tuff ring on left. Trees on the left hide remnants of the double scoria cones. *Sir George Grey Special Collections, Auckland City Libraries*
⊙ Part of a painting by John Kinder in 1861 showing the profiles of Purchas Hill (left) and Mt Wellington from the west. *Sir George Grey Special Collections, Auckland City Libraries*

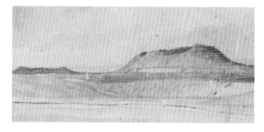

development in the late 2000s, a major new road was built through the footprint of the southern half of Purchas Hill's scoria cone. Today, none of the original shape of the Purchas Hill scoria cone is left. The remaining knob of scoria marks the junction of the two cones. It and the footprint of the northern cone is intended to be set aside as a reserve and maybe one day the northern cone will be rebuilt.

Places of interest around Te Tauoma/Purchas Hill:
See page 196.

⬆ The southern scoria cone of Te Tauoma/Purchas Hill viewed from the northern slopes of Mt Wellington in 1921. *James D. Richardson, Sir George Grey Special Collections, Auckland City Libraries*
⬇ View of the site of Te Tauoma/Purchas Hill Volcano (centre) from Mt Wellington in 2012 when the roads had been constructed but not the dwellings in this part of Stonefields subdivision. A small knob of scoria in the centre of the photo is all that is left of the two quarried-away cones.

Maungarei/
Mt Wellington

⊙ Maungarei/Mt Wellington scoria cone and its triple crater (one filled with a reservoir) from the southeast. The steep, pine-covered slope in the left foreground is the former face of the Mt Wellington Borough scoria quarry that was closed in 1967. The pines were removed in 2019. *Photo by Alastair Jamieson, 2018*

Land status: Mt Wellington Domain is administered by the Tūpuna Maunga Authority.

What to do: Walk up and around the crest of the high scoria cone for spectacular panoramic views. Explore Winifred Huggins Woodlands on the east side.

Geology

Maungarei/Mt Wellington has the tallest scoria cone, about 100 m from base to summit, in the Auckland Volcanic Field. It is also the second-youngest of Auckland's volcanoes, erupting shortly after neighbouring Purchas Hill close to 10,000 years ago. The geological sketch plans prepared by Charles Heaphy in 1857 and Ferdinand von Hochstetter in 1859, before quarrying removed some of the features, form the basis of our modern understanding of the landforms of Purchas Hill and Mt Wellington.

Mt Wellington's scoria cone has a near circular base and rises to a flattish rim (135 m ASL) with three small fountaining craters that overlap with no scoria embankments between them. The northern crater is mostly filled with a concrete water reservoir installed in 1960. As fountaining eruptions proceeded above, a large quantity of lava poured out from vents at the foot of the

○ A heap of fused cowpat bombs thrown out by late phase fiery-explosive eruptions forms a small bluff above the main crater of Maungarei/Mt Wellington.

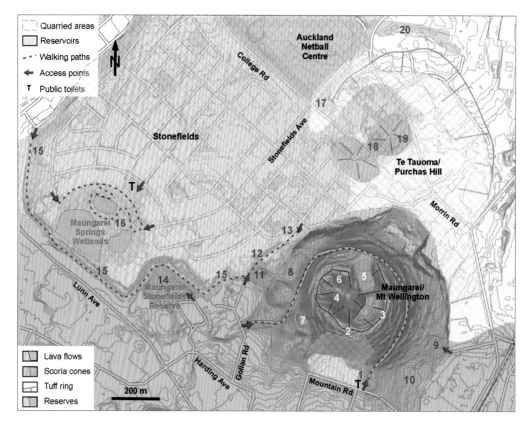

Legend:
- Quarried areas
- Reservoirs
- - - Walking paths
- ← Access points
- T Public toilets

College Rd

Auckland Netball Centre

20

17

Stonefields

Stonefields Ave

18 19

Te Tauoma/
Purchas Hill

15

T

Maungarei 16
Springs
Wetlands

13

12

Morrin Rd

15

14
Maungarei
Stonefields
Reserve

15 11

8

6 5

4

Maungarei/
Mt Wellington

Lunn Ave

7

2

3

9

- Lava flows
- Scoria cones
- Tuff ring
- Reserves

200 m

Harding Ave

Gollan Rd

Mountain Rd

T

10

Places of interest around Maungarei/Mt Wellington and Te Tauoma/Purchas Hill:

1. Main entrance and car park for Maungarei/Mt Wellington Domain.
2. Walk up the sealed road from one of the two entrances and enjoy the view from the top.
3. Walk around the crest of the scoria cone craters with its extensive Māori earthworks.
4. The crater formed by fountaining eruptions from three vents – two still visible.
5. The third crater is now filled with a concrete water reservoir with a flat-topped field on top.
6. Bluff composed of welded pile of cowpat volcanic bombs erupted late in the sequence.
7. Flat-topped knoll on western slopes of cone formed by another buried water reservoir.
8. Sealed concrete top of Te Rua-a-Pōtaka surge chamber. No access – the chamber is wāhi tapu, a sacred site.
9. Winifred Huggins Woodlands planted from 1969 as a memorial to her contribution to the Auckland Tree Society. Mix of exotic deciduous and evergreen specimen trees.
10. Pre-European lava-field gardens and pits on lower slopes of cone.
11. Large depression formed by roof collapse of a lava cave.
12. Goldsbury Track walking path from end of Gollan Rd to Stonefields with exposures of lava flows, some highly frothy, with scoria above.
13. Weathered exposures of remnants of Purchas Hill tuff ring under lava flows and scoria.
14. Maungarei Stonefields Reserve with pre-European gardening stone heaps and unquarried portion of lava-flow levee.
15. Maungarei Heritage Trail runs along old quarry bench cut in a thick lava flow.

16. Maungarei Springs Wetland was the deepest part of the old quarry and now contains three artificial lakes for waterfowl. Pumphouse used to flush excess Stonefields' stormwater via Waiatarua Reserve wetlands to Ōrākei Basin.
17. Weathered road cut of tuff from Purchas Hill tuff ring; beside Stonefields Ave.

18. Remnant knoll of scoria from inside Te Tauoma/ Purchas Hill's double scoria cone, left when quarrying ceased.
19. Derelict concrete structures from quarry and crusher activities.
20. Large blocks of bedded tuff from Purchas Hill tuff ring in Merton Reserve.

northwestern, western and eastern sides of the scoria cone. The exuded lava flows filled the deep valley to the west, with the main stream of lava flowing southwest in a wide path through Ellerslie towards Penrose, where its path was blocked by earlier flows from One Tree Hill and Mt Smart. A small branch of lava flowed around the east side of the Mt Smart flows along the route of present Great South Rd and finally came to a stop in the vicinity of Southdown. Other lobes of lava flowed into the explosion crater of Purchas Hill and over the sandstone ridge in the vicinity of present-day Merton Rd and down towards Glen Innes, where it

stopped near Apirana Ave. A small lava flow also spread southeastwards from a vent somewhere near the entry gate to Mt Wellington Domain; the lava-flow front came to a stop in the vicinity of the modern railway line with a lobe extending under Panmure Roundabout and Ireland Rd.

At the western foot of the scoria cone is the sealed top of Te Rua-a-Pōtaka, a 16-metre-deep, bell-shaped lava shaft that was the surge chamber through which lava swirled as it poured out to feed the extensive lava flows to the west. Other lava caves are also known in Mt Wellington's flows.

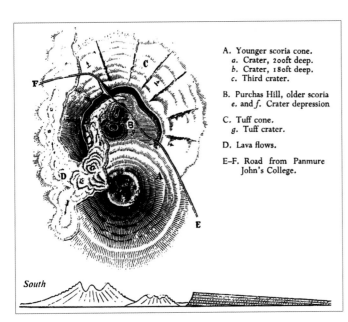

A. Younger scoria cone.
  a. Crater, 200ft deep.
  b. Crater, 180ft deep.
  c. Third crater.

B. Purchas Hill, older scoria
  e. and *f.* Crater depression

C. Tuff cone.
  g. Tuff crater.

D. Lava flows.

E–F. Road from Panmure John's College.

◔ Hochstetter's pencil and ink sketch plan and cross-section of Mt Wellington and Purchas Hill made in 1859. *Courtesy of Hochstetter Private Collection, Basel*

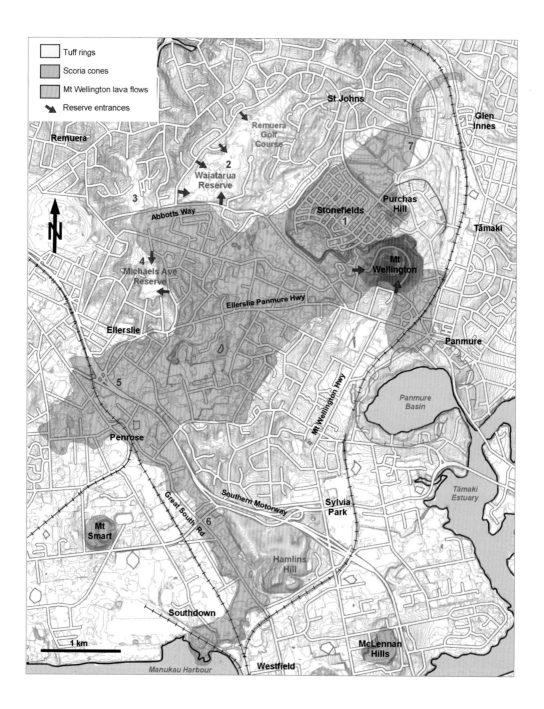

Tuff rings

Scoria cones

Mt Wellington lava flows

Reserve entrances

St Johns

Glen Innes

Remuera

Remuera Golf Course

2
Waiatarua Reserve

3

Abbotts Way

Stonefields
1

Purchas Hill

Tāmaki

N

4
Michaels Ave Reserve

Mt Wellington

Ellerslie Panmure Hwy

Ellerslie

Panmure

Mt Wellington Hwy

Panmure Basin

5

Penrose

Tāmaki Estuary

Great South Rd

6

Southern Motorway

Sylvia Park

Mt Smart

Hamlins Hill

1 km

Southdown

McLennan Hills

Manukau Harbour

Westfield

○ *Places of interest around Maungarei/Mt Wellington lava-flow field:*

1. Former Mt Wellington quarry that excavated thick Mt Wellington basalt lava flows – now Stonefields suburb.
2. Waiatarua Reserve is a partially drained lake/swamp that was dammed when lava flows from Mt Wellington flowed across the mouth of the valley.
3. Koraha Reserve up to College Rifles Park is a drained former swamp formed when lava flows dammed the mouth of the valley.
4. Michaels Ave Reserve is drained swampland dammed between Mt Wellington flows and the sandstone ridge.

5. Southern Motorway road cuttings just south of Ellerslie–Penrose interchange composed of columnar-jointed basalt lava flow from Mt Wellington.
6. Road cuttings beside Great South Rd–Southern Motorway interchange of Mt Wellington lava-flow basalt.
7. Cutting in columnar-jointed basaltic lava flow on pavilion side of Colin Maiden Park cricket ground. This northern lobe flowed down a small valley as far as Glen Innes.

○ Vertical aerial view of Mt Wellington scoria cone in 1954 showing the three-centred crater prior to installation of the water reservoir and prior to construction of the road up it. *Courtesy of LINZ historic aerial photo archive*

◔ Looking across the extensive lava-flow field to Maungarei scoria cone before the basalt rock was quarried away for aggregate and the quarry filled with houses to become the suburb of Stonefields. *Gwyneth Richardson, 1930s, Private collection*
◔ South side of Maungarei/Mt Wellington in 1927 showing the enormous Mt Wellington Bourough Council scoria quarry. *James D. Richardson, Sir George Grey Special Collections, Auckland City Libraries*

Human history

The Māori name Maungarei means 'the watchful mountain' or 'the mountain of Reipae', a Tainui ancestress who travelled to Northland in the form of a bird. Maungarei is an impressive archaeological site. Its flanks, particularly on its eastern side, are covered in terraced house sites and food-storage pits. The crater rim has three strong points each defended by transverse ditches.

The mountain received its European name from colonial surveyor Felton Mathew in honour of the Duke of Wellington. Quarrying of Mt Wellington's scoria cone began on the steep southern slopes in the 1850s and continued on and off until 1967. For 50 years it supplied scoria for the Mt Wellington Roads Board. Pine trees were planted over the recontoured slope soon after quarrying ceased and their removal began in 2018.

Mt Wellington quarry was the largest aggregate quarry in New Zealand for much of the latter part of the 20th century. It was sited in the 25-metre-thick flows that filled the head of the valley to the west of the cone. Heat from these flows had baked the underlying clay on the valley slopes, forming a red-brown natural brick that was often seen in the western and northern walls of the quarry. All quarrying ceased here in 2008 and the giant hole has since been transformed into a new suburban subdivision called Stonefields. A brand new centre for Auckland netball was opened in part of the rehabilitated Mt Wellington quarry in 2007.

◐ Maungarei Springs
Wetland in the new suburb
of Stonefields has been
opened in the Mt Wellington
aggregate quarry. The
building in the foreground
pumps excess stormwater
from the lakes out to Ōrākei
Basin via Waiatarua Reserve.
The path on the right is the
heritage trail that follows a
former quarry bench.

Maungarei Stonefields Reserve and Heritage Trail

Land status: Maungarei Stonefields Reserve (26 Tidey Rd), Mt Wellington Heritage Trail and Maungarei Springs Wetland are all public reserves managed by Auckland Council.

What to do: Visit a small remnant of the extensive stonefields cultivation sites that surrounded Maungarei in pre-European times; walk the Mt Wellington Heritage Trail and loop back through Stonefields subdivision and Maungarei Springs Wetland.

West of Maungarei scoria cone the thick, valley-filling lava flows became the huge Lunn Ave basalt quarry, which in the late 20th century was New Zealand's largest aggregate quarry. Since quarrying finished the huge hole has been transformed into the affluent residential suburb of Stonefields. The high quarry walls of thick basaltic flows remain around the southern and eastern sides of the suburb with a new wetlands reserve named Maungarei Springs the centrepiece along the foot of the southern wall. The created ponds collect all the stormwater from the new suburb, much of which is then pumped away to Waiatarua Reserve and out to Ōrākei Basin.

◑ Highly frothy basalt (pseudo-reticulite) in the top of a lava flow exposed beside Goldsbury Track, which leads down to Stonefields suburb from Maungarei/Mt Wellington cone. Width of photo 20 cm.

201

Maungarei Stonefields archaeological reserve in Tidey Rd, Mt Wellington contains the only remains of the extensive pre-European Māori gardens on the rich volcanic soils developed on Mt Wellington's lava-flow field. The bush-covered ridge on the top right is part of the lava-flow levee left by a passing flow from Mt Wellington. *Photo by Alastair Jamieson, 2018*

A former bench around the quarry wall has been made into a heritage trail with safety fences on both sides. This trail gives excellent views across the former quarry. Exposures of thin flows with gas blisters, lava stalactites and branch moulds can be seen beside the trail in the south-east corner. Goldsbury Track that links Stonefields subdivision with the end of Gollans Rd has rock exposures along its route showing where the basaltic lava oozed out from beneath the scoria cone to flow off down the valley.

A small area on the top of the flows that was saved from quarrying is now the Maungarei Stonefields Reserve and contains the last remaining pre-European cultivation area on the Mt Wellington lava-flow field. There are numerous heaps of stones that were mounded up to create warm oases for growing subtropical crops. Along the south side of the reserve is a large ridge of broken-up basalt that was the levee on one side of a lava flow. Most of the molten lava flowed on, but this ridge of basalt rubble that had cooled and solidified more quickly on the outside of the flow was left behind and shows how thick the flow once was.

Waiatarua Reserve, Meadowbank, from the northeast in 2018. It was formerly a wetland-ephemeral lake dammed by a Mt Wellington lava flow that Abbotts Way now runs along (upper left). *Photo by Alastair Jamieson, 2018*

⊝ View east across Waiatarua/Lake St John in 1917 with Maungarei/ Mt Wellington in the distance. *Auckland Weekly News, Sir George Grey Special Collections, Auckland City Libraries*

Waiatarua and Michaels Ave reserves lava-flow-dammed lake and swamp

What to do: At Waiatarua Reserve you can jog, cycle or walk (40 minutes) around the wetland or take the dog for a run (off-leash area); playground and skateboard bowl at Grand Drive car park. At Michaels Ave Reserve you can jog or walk around the perimeter tracks and exercise trail, feed the birds (100 Michaels Ave) or play ball on the playing fields; playground at 46 Michaels Ave entrance; Ellerslie AFC has artificial turf soccer fields here, the southern fields are home to the Ellerslie Cricket Club and the YMCA has a gym.

Geology

The largest lava flows from Mt Wellington initially flowed west into Stonefields valley and then southwest down a wide shallow valley through Ellerslie. One arm of these flowed back up another branch of the valley towards Remuera, coming to a halt at Abbotts Way and Michaels Ave. These latter flows cut off and dammed a number of side valleys further upstream. Two of these developed into small swamps, now drained and enjoyed as Michaels Ave Reserve, and Koraha Reserve and College Rifles sportsground, but the largest and best known is Waiatarua Reserve in Meadowbank. At Waiatarua, the new basalt dam created a shallow freshwater lake that was still present after heavy rain in early European times when it was named Lake St John. Vegetation grew in the Waiatarua wetland and 3 m of peat accumulated on its floor over the last 10,000 years.

Human history

The Māori name for this ponded lake is Waiatarua, meaning 'the waters of reflection'. Waiatarua Reserve land was mostly gifted to the citizens of Auckland by the owner Mr R. Abbott in 1918. In the 1920s, the council decided that the periodic flooding of 22-hectare Lake St John was a nuisance and so, in 1929, amidst much public opposition, a tunnel was dug through the sandstone under Remuera Rd ridge and a permanent drain created through to Ōrākei Basin. Today, Waiatarua Reserve has a central artificial wetland that is used to trap silt from stormwater that flows into it from surrounding land and is also pumped there from Maungarei Springs in Stonefields. Many native trees and shrubs have been planted and the wetland and ponds are now home to numerous pūkeko, mallard and grey ducks, little shags and white-faced herons. Grass carp were introduced into the waterways in 1994 to help control the waterweeds that clogged the ponds.

Te Kopua Kai-a-Hiku/
Panmure Basin

⬆ View from the north over intertidal Panmure Basin. Sea water flows in and out via a narrow deep channel eroded through the tuff ring, which links to the Tāmaki Estuary on the left. *Photo by Alastair Jamieson, 2009*

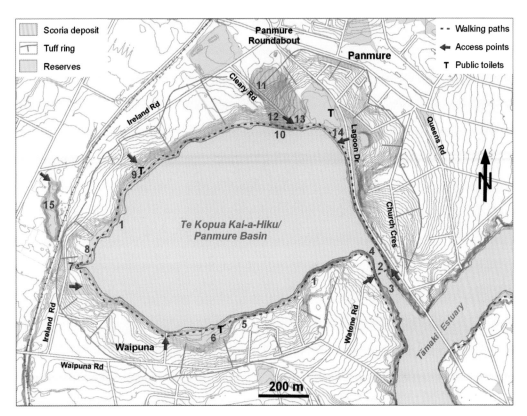

Places of interest around Te Kopua Kai-a-Hiku/Panmure Basin:

1. Path leading right around the inside of Panmure Basin (3 km, 45 minutes).
2. Walking bridge over the breached entrance to Panmure Basin from Tāmaki Estuary.
3. Pied shag colony in trees on southeast side of walking bridge.
4. Layers of bedded and slumped tuff in low cliffs on both sides of basin entrance.
5. Take high path for better view over the basin.
6. Waipuna Miniature Railway (open Sunday afternoons for rides) and Scale Marine Modellers boat pond. Car park.
7. Small walking bridge over inlet stream from van Damme's Lagoon.
8. Flat area beside basin is site of former Ireland Bros Tannery.
9. Young children's playground and Panmure Lagoon Sailing Club. Car park.
10. Beside the round-the-basin path is an exposure of black and red scoria that was erupted by fire-fountaining from the centre of the crater.
11. Cleary Rd hill composed of wind-blown scoria mantling a high point on the tuff ring.
12. Adventure playground for older children, near end of Cleary Rd.
13. Cleary Rd car park at back of Panmure Lagoon Pool and Leisure Centre.
14. Skateboard park and small beach, unsuitable for swimming. Lagoon Drive car park.
15. Van Damme's Lagoon (public access at 112–114 Mt Wellington Highway).

205

◐ Vertical aerial photo of Panmure Basin in 1955 as suburban subdivision was beginning to spread around its northern side. *Courtesy of LINZ historic aerial photo archive*

Land status: Panmure Basin reserves are managed by Auckland Council.

What to do: Walk, jog or ride a bike on the sealed path that goes around the inside of the crater. Also present are exercise stations, skateboard park (Lagoon Drive), playgrounds (Cleary Rd, Ireland Rd car park) and miniature train rides (Peterson Reserve).

Geology

Panmure Basin is a 1.5-kilometre-diameter explosion crater formed by a series of wet eruptions that threw out a large volume of volcanic ash and broken-up Waitematā Sandstone through which it erupted 25,000 years ago. This material built up a 25-metre-high tuff ring that encircles the basin and is now mostly covered in houses, shops and the Waipuna Hotel complex. Layers of bedded ash (tuff) can be seen around the basin shores and in the cliffs on either side of the entrance channel. The tuff beds exposed in these entrance cliffs show that some of the tuff ring's inner slopes slumped back into the crater during the eruptions and the chaotic pile was later buried by further ash layers. Intertidally, on the west side of the basin, thin tuff can be seen draped over a gently rolling pre-volcanic land surface composed of white rhyolitic sediment.

Volcanic ash accumulated on both sides of the Tāmaki River valley and may have temporarily dammed its flow. Previous reports of a buried fossil forest beneath the tuff on the banks of the Tāmaki appear to relate to a much older fossil forest preserved within the underlying 1-million-year-old black peat.

Recent drilling has shown that the wet explosive style of eruptions that formed the basin later switched to dry fire-fountaining. This built a small scoria cone inside the explosion crater, the top of which is now buried beneath the mud-filled floor. The small hill in Cleary Rd near Panmure Roundabout consists of 1–4 m of scoria (wind-blown from fire-fountaining in the basin) mantling a high point on the tuff ring. Similar fine scoria can be seen in a small cutting on the side of the round-the-basin pathway at the end of Cleary Rd.

⊕ Impact depressions from projectile blocks of basalt and Waitematā Sandstone that landed in soft tuff, seen in the cliffs on the west side of the entrance channel into Panmure Basin.

After volcanic activity ceased, the crater filled with fresh water, creating a lake. Overflowing water eroded down the low northeastern corner of the tuff ring and rising sea level after the Last Ice Age caused the sea to invade the crater at this point, which has been a tidal lagoon for the last 8000 years.

Human history

The traditional name for this feature is Te Kopua Kai-a-Hiku, meaning 'the eating place of the guardian taniwha Moko-ika-hiku-waru', which is also the origin of the name Mokoia, the former Māori village located at the basin's northeastern entrance. On its western shores was the sacred spring Te Waipuna-a-Rangiātea, named by the *Tainui* crew after a spring in East Polynesia. In 1821, Mokoia pā was occupied by many hundreds of Ngāti Pāoa people when it was besieged by a musket-wielding Ngā Puhi war party who killed most of them.

The European name Panmure Basin was given by Governor Sir George Grey after the Hon. Maule Panmure, secretary of war at that time back home in England. From the 1860s to

1923, the Ireland Bros Tannery operated from a site in the southwest corner of Panmure Basin. Across the railway line from here is Van Damme's Lagoon, a small reserve containing a wetland ponded behind a dam built in the 1860s for the tannery. After the tannery closed, the lagoon area was owned by Theodor van Damme, who set about beautifying it with numerous plantings and paths. It passed into public ownership in 1975.

Today, the Auckland Society for Model Engineers has a clubhouse, miniature railway line and model boat pond alongside Panmure Basin in Peterson Reserve and the Panmure Lagoon Sailing Club has its clubhouse and boat lockers beside the Ireland Rd car park.

Lagoon biota

Mangroves that dot the shoreline of the basin are remnants of a once more extensive mangrove forest margin. Bird life is plentiful around the tidal flats with ducks, gulls, white-faced herons, kingfishers, and little and black shags. A small colony of pied shags can be seen nesting in the trees beside the entrance channel.

⊕ View south along the west side of Panmure Basin about 1910 across the embayment that now has the Ireland Rd car park and Panmure Lagoon Sailing Clubhouse. Ireland Bros Tannery buildings are on the edge of the basin on the left. *James D. Richardson, Sir George Grey Special Collections, Auckland City Libraries*

Ōhuiarangi/
Pigeon Mountain

⊙ View from the west over the original extent of Pigeon Mountain tuff ring and explosion crater. Half Moon Rise (left) runs along the northern crest of the tuff ring. Britannia Place (centre) is on the floor of the quarried-out crater and Pakuranga Domain (right) contains the remaining half of the volcano's scoria cone with its pre-European terracing visible. *Photo by Alastair Jamieson, 2009*

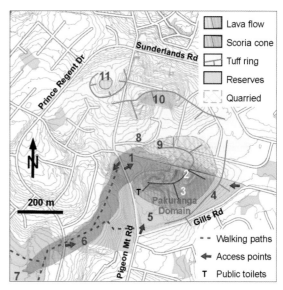

Places of interest around Ōhuiarangi/Pigeon Mountain:

1. Captain Musick Air Scouts clubhouse and Pigeon Mountain Kindergarten. Car park.
2. Track leads through old breached crater then up to the summit of remaining part of scoria cone (53 m ASL). Note shell midden on the track on the way up and earthwork remains of pre-European pā on summit. Don't confuse an area of quarrying near the summit with Māori earthworks.
3. Pre-European kūmara pits – the lower parts of buildings used for food storage.
4. Southeastern slopes of cone have subtle boundary drains between former pre-European gardening areas.
5. Sports fields.
6. Children's playground and track down valley following route of lava flow.
7. Wakaaranga Creek Reserve with remnants of the toe of a lava flow in salt marsh around its head. Start of Rotary Walkway and cycle path along the edge of Tāmaki Estuary to Panmure bridge.
8. Location of Pigeon Mountain explosion crater and Howick District's water pumping station (near 14 Britannia Place).
9. Walkway between 21 and 23 Britannia Place to Glennandrew Drive; in footprint of quarried-away half of scoria cone.
10. Slopes in Half Moon Park (between ends of Britannia Place and Half Moon Rise) are part of the inner wall of the explosion crater with the crest of the tuff ring at the top.
11. Site of former small explosion crater in private reserve – no public access.

◔ Only half of Pigeon Mountain scoria cone remains; the other half was removed by quarrying in the 1950s–70s. Terracing from the pre-European pā is clearly visible on the upper parts. *Photo by Jill Kenny*

Land status: Pakuranga Domain is administered by the Tūpuna Maunga Authority. Half Moon Park and Wakaaranga Creek Reserve are also public reserves. Other parts of the tuff ring and crater floor are in private properties. The route of the lava flow down to Wakaaranga Creek is also in reserve land.

What to do: Walk to the top of the cone for the view and see the pre-European earthworks. Cross Pigeon Mountain Rd and walk down the track to the playground or continue on foot or cycle along the Rotary Walkway beside the Tāmaki Estuary.

Geology

Pigeon Mountain was formed by volcanic eruptions 24,000 years ago. It began with wet explosive eruptions that built a 500-metre-wide crater with surrounding tuff ring. The higher northern and eastern crest of the tuff ring is still recognisable beneath suburban houses on Half Moon Rise. A scoria cone was built by fountaining eruptions inside the explosion crater. As the cone was being formed, a small lava flow poured out from the breached crater on its northwestern side and flowed 700 m southwest down the valley, coming to a halt at the head of present-day Wakaaranga Creek, beyond present-day Prince Regent Drive. A small, 50-metre-diameter explosion crater erupted through the tuff ring near its northwestern crest.

Human history

The Māori name for this volcano is Ōhuiarangi, meaning 'the desire of Rangi and Pakuranga'. Some of the defensive earthworks of the pre-European pā still exist on the scoria cone remnant. Long, shallow boundary drains that separated Māori gardening areas are just visible

◐ Vertical photo of Ōhuiarangi/
Pigeon Mountain Volcano in
1940 prior to the period of severe
quarrying. All the cone north of
the major boundary fence has now
been removed. A small circular
explosion crater can be seen clearly
in the northwestern sector of the
tuff ring. A narrow lava flow extends
down the valley from the breached
(U-shaped) scoria cone to the
head of Wakaaranga Creek (bottom
left). *Courtesy of LINZ historic aerial
photo archive*

in certain light where they run down the eastern
grassed slopes of the cone towards Gills Rd.

The European name Pigeon Mountain comes
from the numerous native pigeons, or kererū,
that lived in the area in early European times. The
southern half of Pigeon Mountain was gazetted
as Pakuranga Domain under the control of a local
domain board in 1881. The northern half was
privately owned and small-scale quarrying began
there in 1913. The scale of quarrying escalated in
the 1950s after the closure of the Mt Wellington
scoria quarry as suburban subdivisions started to
extend east of the Tāmaki Estuary. After a major
campaign to stop the quarrying, it finally ceased
in the 1970s but half the cone had already gone.

A shallow well and bore on the northwest foot
of Pigeon Mountain was used by Howick Borough
and nearby Bucklands and Eastern beaches as a
water supply from 1930. This was phased out in
the 1950s as reticulation links to Auckland and
the Hunua Ranges were installed.

◔ Quarrying of Pigeon Mountain scoria cone was well
advanced by 1970. *Howick Library*

East Tāmaki Volcanoes

○ View south over the line of four East Tāmaki volcano sites (circled) in 2009. They are from north (front) to south (rear): Styaks Swamp, Green Mount, Ōtara Hill and Hampton Park. *Photo by Alastair Jamieson*
○ Map of the original extent of tuff rings, scoria cones and lava flows of the four East Tāmaki volcanoes – Styaks Swamp, Green Mount, Ōtara Hill and Hampton Park.

A north–south line of four volcanoes now lies within the East Tāmaki industrial area, between Harris and Springs roads and Te Irirangi Drive. These four volcanoes were fed by magma that rose up along a fault line in the deeply buried basement greywackes. Volcanic activity appears to have migrated from south to north with Hampton Park being the oldest and Styaks Swamp the youngest. All four centres began their eruptions with wet explosive activity that threw out considerable volumes of ash. Dominant westerly winds during these eruptions resulted in the surrounding tuff rings building up much higher on the downwind eastern sides of all the explosion craters. Subsequent dry explosive eruptions built scoria cones at the southern three volcanoes. All lava flows spread westwards after overtopping the low tuff rings on their upwind side.

○ *Places of interest around the East Tāmaki volcanoes:*

1. Location of Styaks Swamp explosion crater, at intersection of Greenmount Drive and Polaris Place.
2. Old drystone basalt wall along Harris and Smales roads marks boundary of quarried-away Green Mount Volcano, resurrected as a landfill hill and soon to be converted into a reserve called Matanginui/Styak-Lushington Park.
3. Greenmount Reserve playing fields with old drystone basalt wall along boundary with site of Green Mount Volcano and extending down the centre of Guys Rd.
4. Site of Ōtara Hill scoria cone in middle of former explosion crater, now covered in factories in the block surrounded by Lady Ruby Drive and Sir William Ave.
5. Accent Drive cuts through the southeast corner of Ōtara Hill tuff ring remnant.
6. Te Puke o Tara Hampton Park (see map on page 221).
7. Cryers Jetty on edge of Pakuranga Creek, made of basalt blocks from Green Mount flows in the 1890s.
8. Robina Place jetty and small basalt quarry, Pakuranga Creek, at high tide in small bay.

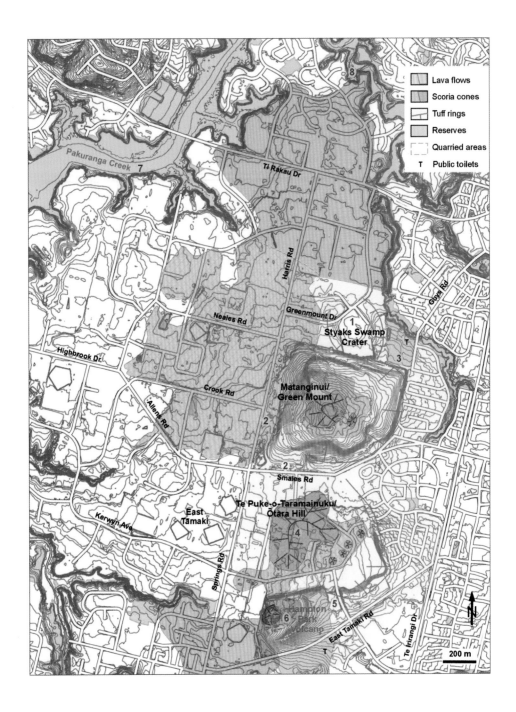

Lava flows
Scoria cones
Tuff rings
Reserves
Quarried areas
T   Public toilets

Pakuranga Creek 7

Ti Rākau Dr

Harris Rd

Neales Rd

Greenmount Dr

Glys Rd

Highbrook Dr.

1
Styaks Swamp
Crater
T
3

Crook Rd

Matanginui/
Green Mount

Allens Rd

2

2
Smales Rd

Te Puke-o-Taramainuku/
Ōtara Hill

East
Tāmaki

4

Kerwyn Ave

Spring Rd

Hampton
6 Park
Volcano

5

East Tāmaki Rd

Te Irirangi Dr

T

N

200 m

Styaks Swamp
Crater

↑ Vertical aerial photo in 1940 showing the drained swampy depression of Styaks Swamp explosion crater (top right) and Green Mount's scoria cone (bottom right) and lava-flow field (left) prior to industrial subdivision. *Courtesy of LINZ historic aerial photo archive*
↓ The site of Styaks Swamp explosion crater and tuff ring is now hardly recognisable beneath the industrial subdivision of Greenmount Drive in the foreground. Beyond is the grassed landfill mound on the site of quarried-away Green Mount Volcano. Greenmount Reserve is the flat sports fields on the left. *Photo by Alastair Jamieson, 2009*

Land status: The site of this small explosion crater is the junction of Greenmount Drive and Polaris Place. It extended under some of the surrounding factories.

The northernmost and smallest of the line of four East Tāmaki volcanoes, Styaks Swamp Crater was named after an early European landowner, John Styak. It erupted about 20,000 years ago, soon after Green Mount to the south. The short-lived, pulsating wet explosive eruptions built a low-profile tuff ring around a 250-metre-wide crater with volcanic ash mantling the Green Mount lava flows that surround it to the north and west. Unlike the other three East Tāmaki volcanoes, there were no dry fountaining eruptions or lava flows.

Core samples from the drained swamp prior to the construction of factories showed that the crater was originally about 30 m deep and soon after formation it filled up with rainwater, becoming a small freshwater lake with a small overflow creek across the lowest rim of the tuff ring to the north. Over thousands of years the lake filled with fine-grained, diatom-rich sediment, eventually shallowing to a vegetated swamp. This swamp was still present and clearly recognisable prior to the industrial subdivision of Greenmount Drive in the 1980s.

Today, the site of Styaks Swamp explosion crater is difficult to identify as it has been modified and partly filled during preparation for industrial development. Soon after installation, a large electricity transformer on this corner reputedly subsided by about 2 m as a result of compaction of the underlying lake sediment and reclamation earthfill.

Places of interest around Styaks Swamp Crater:
See pages 212–13.

215

Matanginui/ Green Mount

Land status: The high landfill on the site of Green Mount Volcano is currently (in 2019) not publicly accessible, but Auckland Council has plans to develop it into a public reserve.

Geology

Green Mount erupted about 20,000 years ago. It began with wet explosive eruptions that built a surrounding tuff ring with the highest part forming the eastern arc, the gentle outer slopes of which still remain. Inside this wide crater, subsequent fountaining built a 60-metre-high main scoria cone capped by a small central crater and a smaller scoria mound to the east. Lava emanating from around the base of the scoria cone initially flooded the moat between it and the tuff ring, before spilling north and west. Molten lava spread over a wide area of low-lying ground, extending 2 km to the banks of Pakuranga Creek, north of present-day Ti Rakau Drive. The sheet of flows spread 1 km

↑ View southwest over Green Mount scoria cone (right) and Ōtara Hill scoria cone (left) in 1949, prior to substantial damage by quarrying. *Whites Aviation, University of Auckland*

○ The quarried-away footprint of Green Mount volcano has been replaced by this low mound made of dirt-covered rubbish, soon to be opened as a public reserve. View from the northeast, 2018.

to the west of Green Mount, its rocky surface defining the location of Allens and Cryers roads, which skirt around its perimeter. Extensive areas of these lava flows were removed by quarrying prior to construction of the industrial subdivision that now hides most traces of them. During this work several lava caves were encountered in the lava flows.

Human history

The meaning of the Māori name, Matanginui, is 'big wind' or 'breeze', referring to the strong air currents that sometimes sweep the cone. The earliest European name for Green Mount was Bessy Bell. In the late 19th century, the land was farmed by John Styak. It was inherited by his daughter Sarah, who married Charles Lushington, and they continued to farm it up until their deaths. The 40-hectare volcano was bequeathed to the former Manukau City Council by Sarah Lushington in 1932 with her wish that it become a public park and recreation ground named Styak-Lushington Park.

Minor quarrying of Green Mount's scoria cone began around 1870 but did not start in earnest until 1964. By 1990 it had been completely removed. Also quarried away has been the basaltic lava in the moat. This quarry subsequently became a large landfill, and now a new grass-covered, gently sloping hill stands in place of the former Green Mount Volcano and fills the entire depression that was once the explosion crater. From the summit of the new hill there are sweeping views across South Auckland to the Manukau and Waitematā harbours. Auckland Council is planning to plant this hill in forest and establish a children's playground, cycleways, walkways and a fitness area.

Places of interest around Matanginui/Green Mount: See pages 212–13.

○ Oblique aerial view from the east over Green Mount in 1958 with quarrying already begun on the main scoria cone. Harris Rd is beyond. *Whites Aviation, University of Auckland*

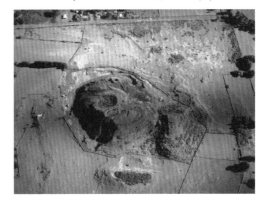

Te Puke-o-Taramainuku/
Ōtara Hill

◔ View northwest in 1949 over Ōtara Hill scoria cone (left) with its breached U-shaped crater and mounds of scoria below that were rafted there on the escaping lava flow. The partly quarried west side of Green Mount cone is on the right and Mt Wellington on the skyline. *Whites Aviation, University of Auckland*
◕ Looking north over the quarried site of Ōtara Hill scoria cone – now a flat area covered by factories, 2009. Some of the eastern tuff ring arc is still present with industry inside the former crater and housing on the gentle eastern outside slopes (to the right). Hampton Park is in front and Green Mount beyond. *Photo by Alastair Jamieson*

Land status: The site of Ōtara Hill Volcano is completely beneath industrial developments and private houses (eastern slopes of tuff ring). The only access is on public roads.

Geology
Like Green Mount, the initial explosive eruptions of Ōtara Hill Volcano built a tuff ring around a 500-metre-wide crater. Westerly winds resulted in most of the ash accumulating as a curved eastern tuff ring and this still exists. It passes southwards into the Hampton Park tuff ring. Dry fountaining eruptions built a large, 60-metre-high scoria cone in the middle of the explosion crater, leaving a moat between the main cone and the tuff ring. Lava flowed from the southeast base of the scoria cone, creating a U-shaped breached crater. This lava rafted parts of the cone out into the moat where they came to rest as four small scoria mounds. The lava partly filled the moat and spilled southwestwards into the explosion crater of adjacent Hampton Park. The age of this volcano and adjacent Hampton Park is currently unknown.

Human history
The modern name Ōtara is derived from the full Māori name of Te Puke-o-Taramainuku, meaning 'the hill of Taramainuku', a Tainui ancestor. Earlier English names included Mary Gray and Smales Mountain, after 19th-century owner the Rev. Gideon Smales. Quarrying began on Ōtara Hill in 1955 and by 2002 all trace of the scoria cones, mounds and basaltic lava had been removed, mostly by W. Stevenson and Sons. Two of the main roads in the subsequent flat industrial subdivision have been named Sir William Ave and Lady Ruby Drive after Sir William Stevenson and his wife.

Places of interest around Te Puke-o-Taramainuku/
Ōtara Hill:
See pages 212–13.

Hampton Park
Volcano

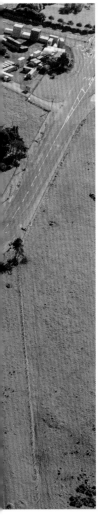

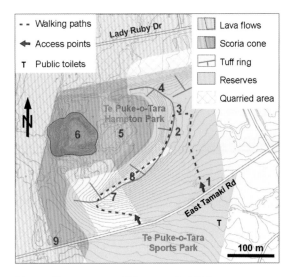

The remains of Hampton Park Volcano, viewed from the south in 2009. The scoria cone sits in the top left corner of the reserve in the middle of the explosion crater, partly surrounded by the tuff ring arc with church and homestead on it (right). *Photo by Alastair Jamieson*

Places of interest around Hampton Park Volcano:

1. Walking entrance to Te Puke o Tara Hampton Park public reserve.
2. Hampton Homestead, home of Gideon Smales; rebuilt after fire in 1940. Private.
3. Remains of local basalt stone stables, severely damaged by fire soon after construction in 1870s.
4. Panoramic views over Auckland and many other volcanic cones.
5. Sunken stone garden. Mostly stone walls but can you find the one exposure of in situ basaltic lava flow?
6. Partly quarried remains of Hampton Park scoria cone in centre of explosion crater. Note remains of small Māori pā with terraces and kūmara pits.
7. St John's Church made of local basalt stone.
8. Path of remembrance runs along crest of part of tuff ring; large kūmara pit.
9. Exposure of lava-flow basalt beneath drystone wall in road cutting on corner.

Land status: Most of what remains of this volcano is in Te Puke-o-Tara Hampton Park, managed by Auckland Council. One part of the tuff ring is publicly accessible church land; Hampton Homestead is leased and not accessible.

What to do: Walk around the farmed park and enjoy reading about the site's history on park noticeboards (2 hours).

Geology

The earliest eruptions from Hampton Park Volcano created a 350-metre-wide explosion crater surrounded by a tuff ring. The highest part of this tuff ring remains as an arc around the south and east sides of the crater. Later, fiery explosive eruptions constructed a small, 12-metre-high partly welded scoria cone in the centre of the explosion crater. The cone is unusual in the large amount of lava it contains, presumably fountaining out and landing on the growing hill while still molten. Quarrying in the 19th century removed the heart of the cone but its lower slopes, with Māori terracing, remain and give a good indication of its former location and size. The original height may have been little more than present, with the small cone capped by a crater.

Lava flowed out from around the base of the cone, partly filling the moat between the cone and tuff ring before overtopping the low tuff ring to the west and spreading out as a double-lobed sheet over the low-lying land. A lava flow from adjacent Ōtara Hill later overtopped Hampton Park's low tuff ring in the north and flowed around the west side of the cone. In the 1850s, European visitors Charles Heaphy and Ferdinand von Hochstetter considered the small Hampton Park Volcano to be part of Ōtara Hill and they did not recognise it as a separate volcano.

Human history

There is no known Māori name specifically for this small volcano, although it was clearly occupied and defended as a pā. The volcano gets the name Hampton Park (earlier name Church Cone) from the name of the homestead of its 19th-century owner, the Rev. Gideon Smales. Smales immigrated to New Zealand as a Wesleyan missionary in 1840. On leaving the service he purchased this land, which included neighbouring Ōtara Hill, in 1851 and established a thriving farm and homestead. The property remained in his family until it was bequeathed to the Salvation Army in 1971, and later sold to Manukau City for a reserve.

The little St John's basalt church and the Rev. Gideon Smales' homestead and basalt stables were built on the crest of the tuff ring. The original Hampton Homestead, built in the 1860s, was burnt down in 1940 and replaced by the present building. The stables were also destroyed by fire, in the 1870s, but some of the stone walls are still standing. Remains of a sunken rock garden constructed by Smales' gardener 'Old Campbell' lie beneath trees in the bottom of the explosion crater east of the small cone.

Basaltic lava from the flow that partly filled the floor of the crater was used for the construction of the church, stables and sunken gardens. Loose basalt on the surface of the flows was cleared for farming and used by Yorkshire stonemason James Stewart to build 8 km of stone walls between the paddocks on Smales' farm.

◔ View west across Hampton Park explosion crater from the eastern tuff ring arc. The floor of the crater under the trees is underlain by basaltic lava, much of which was used in the construction of a sunken garden in the late 19th century.
❍ Vertical aerial photo of Hampton Park Volcano showing extent of lava flows out to the west prior to nearby industrial subdivision, 1940. *Courtesy of LINZ historic aerial photo archive*
❍ Remains of the stone stables built from local basalt in the 1870s.

Pukewairiki/
Highbrook Park

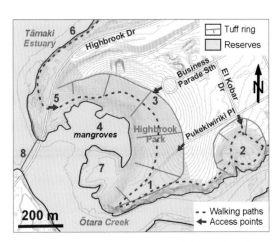

Places of interest around Pukewairiki/Highbrook Park:

1. Main car park.
2. Small satellite crater, now slightly modified as a stormwater pond.
3. Walk the crest of the tuff ring with great views.
4. Coastal wetland in floor of breached crater.
5. Highbrook Drive walkway entrance to Pukewairiki crater.
6. Walkway/cycleway alongside Tāmaki Estuary.
7. Flat land inside the crater is inferred to be eroded remnant of intertidal sediment that accumulated in the crater when sea level was higher, about 125,000 years ago.
8. Ōtara Creek weir.

⊙ Pukewairiki explosion crater from the southwest in 2018. The low south side of the tuff ring has been breached by the sea and removed by erosion. The blue sediment pond (far right) has been constructed in a small satellite crater.
Photo by Alastair Jamieson

Land status: The inside of the main explosion crater and smaller satellite crater lie within a large public reserve managed by the Highbrook Park Trust.

What to do: Enjoy the views from the crest of the tuff ring; walk, cycle or jog some of the walkways (total length 12 km) around the volcano and alongside the banks of the Tāmaki Estuary.

Geology

Pukewairiki Volcano consists of a large explosion crater partly surrounded to the north and east by a high tuff ring mound. A second, much smaller satellite explosion crater is located 500 m to the east on the banks of Ōtara Creek. It has been modified and dammed to create a stormwater pond but also lies within Highbrook Park. Geophysical studies by graduate student Sian France show that solid basalt lies at a shallow depth beneath the floor of Pukewairiki's main explosion crater. This is probably a solidified lava lake or ponded lava flows that partly filled the crater during the last phase of eruption. A precise age for Pukewairiki's eruptions is not known, but it is considerably eroded and appears to be one of Auckland's older volcanoes. Along the edge of the Tāmaki Estuary there is a terrace 10 m ASL. It was probably eroded out of the side of the tuff ring by the sea when sea level was higher than present during the Last Interglacial period, about 125,000 years ago. Tectonic uplift may well have contributed to the present height of the terrace. It was probably at this time that the sea also topped the lowest part of the tuff ring, breaching and starting to erode the south side of the explosion crater.

A small satellite explosion crater on the outer eastern slopes of Pukewairiki tuff ring has now been converted into a stormwater pond.

In 1958, Pukewairiki explosion crater and tuff ring (centre) were part of Ra Ora Stud and the major residential suburbs of Ōtara and Pakuranga had yet to be built over the surrounding farmland. The Southern Motorway across the Tāmaki Estuary was still being constructed. *Whites Aviation, University of Auckland*

Human history

This volcano's Māori name, Pukewairiki, aptly means 'the hill with the associated small lagoon'. For some time the crater has been wrongly known as Pukekiwiriki, which more correctly applies to the remnant older volcano at Red Hill, 10 km to the south at Papakura. The volcano has no European name and was not recognised by Europeans as a volcano until the 1950s.

In 2005, Sir Woolf Fisher's Ra Ora racehorse stud farm became Highbrook Business Park and, in 2007, Pukewairiki crater was opened to the public as a reserve within the developing industrial subdivision. At the same time, Highbrook Drive was built across the mouth of Ōtara Creek and around the western side of the Pukewairiki tuff ring to link Highbrook and East Tāmaki with the Southern Motorway at Ōtāhuhu. Today the crater floor is mostly filled with salt marsh and mangrove forest. Sea water level never drops far during the tidal cycle because a high weir has been built across the mouth of Ōtara Creek. This provided a supply of cooling water for the now decommissioned Ōtāhuhu gas-fired power station, which stood on the southern bank opposite the park.

Te Apunga-o-Tainui/ McLennan Hills

Land status: The footprint of the former cones (Richmond Precinct) and most of the shield volcano lies in privately owned residential and industrial properties. A public esplanade reserve runs around the Tāmaki Estuary shoreline of the eastern lava flows.

Geology

McLennan Hills Volcano consisted of four small scoria cones, each with a central crater that erupted in the middle of an explosion crater with surrounding tuff ring. Lava poured out from the base of the cones, filling the explosion crater and burying most of the tuff ring as it flowed out to the north, east and south, forming a small shield volcano. The flows underlie all the Panama Rd peninsula and 'Tip Top Corner' on the Southern Motorway. The toe of the flows forms the coast of the Tāmaki Estuary from Ōtāhuhu north to the

⊙ Te Apunga-o-Tainui/McLennan Hills from Mt Richmond in 1861. In the foreground is the Ōtāhuhu military camp occupied by the 70th regiment at the time. *G. H. Cooper, Auckland Art Gallery*

⊙ Te Apunga-o-Tainui/McLennan Hills scoria cones from the northeast in 1949, prior to the quarrying that removed them. *Whites Aviation Collection, Alexander Turnbull Library*

⊙ Te Apunga-o-Tainui/McLennan Hills shield volcano from the southwest. In 2018, the footprint of the quarried-away scoria cones was being developed for high-density residential living. Panama Rd peninsula beyond is made of lava flows from this volcano. *Photo by Alastair Jamieson*

⊙ Te Apunga-o-Tainui/McLennan
Hills shield volcano from the northeast
in 1949. The eastern lava-flow apron
in the foreground is now covered in
residential houses on Panama Rd
peninsula. *Whites Aviation Collection,
Alexander Turnbull Library*

Panmure Basin tuff ring near Waipuna Rd bridge.
To the north the flows reached the southern base
of Hamlin Hill sandstone ridge.

Most of the northwestern lava flows and
scoria cones of McLennan Hills are mantled by
ash from neighbouring Mt Richmond, indicating
that McLennan Hills, which has been dated
by the argon-argon method at about 40,000
years, erupted before Mt Richmond. These two
volcanoes, together with Mt Robertson, blocked
the Tāmaki River, which up until that time
flowed southwest into the Māngere branch of
the Manukau River system. The blocked Tāmaki
River overtopped the sandstone ridge between
Glendowie and Half Moon Bay and began flowing
northwards in the opposite direction, eroding
down its course.

Human history

The original Māori name is Te Apunga-o-Tainui,
which may refer to the bow wave or prow of
the *Tainui* canoe, or the point where the canoe
landed about 600 years ago. McLennan Hills was
named after European landowner and Member
of Parliament Ewan McLennan (1861–1948).
In the early 20th century, Hellaby Freezing
Works drew water from a bore in McLennan
Hills' scoria cones and the Westfield Freezing
Works were supplied from groundwater in
the northwestern flows. Apart from a small
quarry on the northeastern cone, the hills were
untouched until 1952. In the 1950s–60s the
scoria cones were completely quarried away.
For several decades the flat platform was used
for greenhouses but at the time of writing the
cones' footprint is the site of construction of 600
terrace houses and apartments in a development
precinct inappropriately named Richmond.

*Places of interest around Te Apunga-o-Tainui/
McLennan Hills:*
See page opposite.

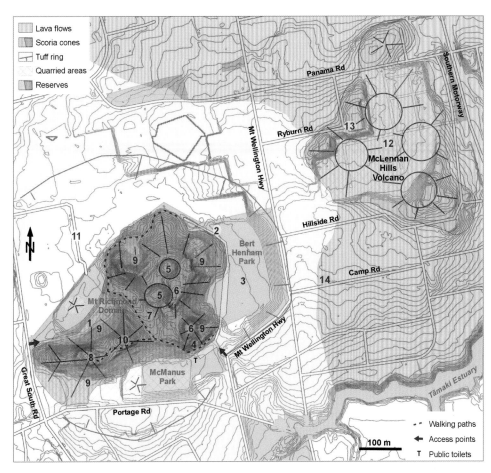

Places of interest around Te Apunga-o-Tainui/McLennan Hills and Ōtāhuhu/Mt Richmond:

1. Western car park and Northern Sports Car Club hall in old quarry.
2. Eastern car park in Bert Henham Park; small playground and Otahuhu Rovers Rugby League Club.
3. Part of explosion crater floor with inner slopes of tuff ring extending up to Mt Wellington Highway.
4. Thin vertical dikes of basaltic lava intrude scoria in cutting beside main path (former road).
5. Two small fountaining craters.
6. Terraced summits of several scoria cones.
7. Derelict historic basalt toilet block under trees.
8. Remaining water reservoir. Site of former reservoir tower.
9. Former scoria quarries.
10. Mt Richmond trig station and views.
11. Reclaimed swampy floor of Mt Richmond explosion crater, now industrial development.
12. Flat-floored old quarry that removed McLennan Hills scoria cones, now site of Richmond housing development.
13. Ryburn Rd. Highest remaining part of McLennan Hills scoria cones.
14. Camp Rd. Site of British forces camp during 1860s New Zealand Wars.

Ōtāhuhu/
Mt Richmond

⊙ View from the south over the remains of Mt Richmond scoria cones and the surrounding explosion crater 'moat' in 2018.
Photo by Alastair Jamieson

○ One of the small fountaining craters on top of the Mt Richmond scoria cones.

Land status: The scoria cones and some of the crater are located within Mt Richmond Domain and Bert Henham Park, administered by the Tūpuna Maunga Authority.

What to do: Explore the scoria cone complex and craters; have a picnic.

Geology

Mt Richmond consists of the partly quarried remains of several scoria cones formed by fountaining from a number of vents. The cones sit in the middle of an 800-metre-diameter explosion crater and surrounding tuff ring, with its lowest rim to the west. In places it is difficult to recognise the original shape of the scoria cones and tuff ring because substantial parts have been removed. The sites of four scoria quarries lie within Mt Richmond Domain and can easily be confused with the two small circular craters in the middle of the complex. Much of the northern part of the tuff ring was flattened during industrial subdivision.

Mt Richmond erupted about 30,000 years ago, blasting its way through the southwestern side of the tuff ring and lava flows of McLennan Hills. Following cessation of volcanic activity, the moat between Mt Richmond scoria cones and the tuff ring became a swamp that over the years partly filled with dark peat.

Human history

This volcano's Māori name Ōtāhuhu is an abbreviation of 'Te Tahuhutanga o Te Waka Tainui', translated as 'the ridgepole of the *Tainui* canoe', referring to the portage of the *Tainui* canoe from the Waitematā to the Manukau Harbour just to the south in the 14th century. All of the scoria cones were extensively modified with surface earthworks for occupation sites and defensive pā in pre-European times, and there are numerous large

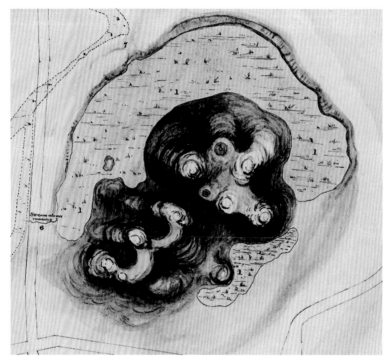

● Plan view of Ōtāhuhu/
Mt Richmond's scoria cone
complex surrounded by
the swampy floor of the
explosion crater. Painted
by Charles Heaphy for
Hochstetter in 1859.
*Courtesy of Hochstetter
Private Collection, Basel*
● Mt Richmond from
McLennan Hills, 1861.
*John Kinder, Auckland
Public Art Gallery*

and small kūmara storage pits. Particularly impressive is the high, steep-sided defensive position at the eastern end of the complex. The mountain was of strategic importance as it commanded the main Waitematā–Manukau canoe portage.

Its European name comes from Major Richmond, a friend of Governor Sir George Grey's in the 1850s. An earlier name was Mt Halswell, after a New Zealand Company commissioner in the 1840s. Mt Richmond Domain was gazetted in 1890. In 1912, the growing Ōtāhuhu Borough established a water supply from a shallow well in the scoria within Mt Richmond explosion crater. By the 1930s, Ōtāhuhu water was supplemented from Auckland City's bulk supply and this local volcanic source was abandoned in 1953. Numerous wild European olives grow on the scoria cones.

◐ Oblique aerial photo of Ōtāhuhu/Mt Richmond from the southwest in 1951, before most quarrying and subdivision, with McLennan Hills cones to the right. *Whites Aviation Collection, Alexander Turnbull Library*
◑ The two-storey-high water tower on the west summit of Mt Richmond in the mid-1950s. It was erected in 1912 and removed in 1961. *Jack Golson, University of Auckland Library*

Mt Robertson/
Sturges Park

◉ View from the southwest of Mt Robertson Volcano. The roads in the bottom and top right run along the crest of the tuff ring enclosing the explosion crater (Otahuhu College playing fields). The scoria cone with the crater excavated out to form a sports amphitheatre lies beyond in the middle of the large explosion crater. *Photo by Alastair Jamieson, 2009*

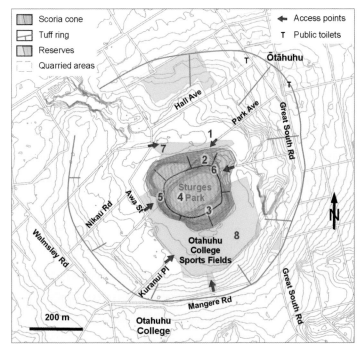

Places of interest around Mt Robertson/Sturges Park:

1. Sturges Park main entrance.
2. Memorial to park donor, Alfred Sturges.
3. Large car park on south rim of scoria cone.
4. Stadium constructed inside the enlarged scoria cone crater.
5. Otahuhu Rugby Football Club building and car park.
6. Otahuhu Softball Club building.
7. Otahuhu and District Highland Pipe Band building.
8. Otahuhu College playing fields in explosion crater 'moat'.

Land status: The scoria cone remnant and most of the floor of the explosion crater are in Sturges Park. The inner tuff ring slopes are privately owned. The crest of the tuff ring underlies parts of Great South and Mangere roads.

Geology

Mt Robertson, in the centre of Ōtāhuhu, is one of the less conspicuous of Auckland's volcanoes. It is a classic castle-and-moat volcano with a simple scoria cone in the centre of an 800-metre-diameter explosion crater with a 12-metre-high tuff ring arc around the south and east sides. Parts of Great South and Mangere roads run along the tuff ring crest, and Otahuhu College is built on the gentle outer slopes of the tuff ring to the south. The low scoria cone was a maximum of 28 m high and had a 200-metre-wide shallow crater. Both the scoria cone crater and the moat between the cone and tuff ring were freshwater swamps that partly filled with accumulated peat and silt in the time since the volcano's eruption 24,000 years ago.

Human history

The pre-European Māori name for this volcano and pā are not known. Prior to the New Zealand Wars this locality was known as Fort Richards, after Lieutenant Richards who set up a defensive

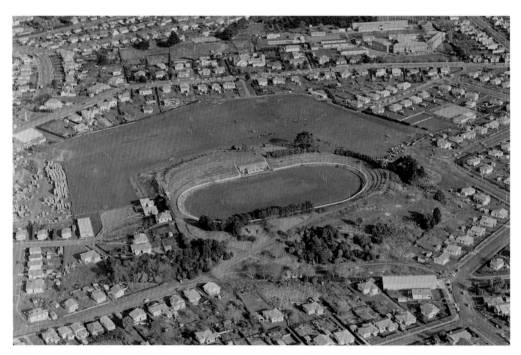

◐ View from the north across Mt Robertson explosion crater and central scoria cone, with Sturges Park stadium in the enlarged crater, 1959. *Whites Aviation Collection, Alexander Turnbull Library*

⊙ The sports amphitheatre in Sturges Park in 2009.
It was created by enlarging the crater in the Mt Robertson
scoria cone.
⊙ Mt Robertson lies within Sturges Park, named after
the donor of the land, Alfred Sturges.

THIS MONUMENT
WAS ERECTED TO
THE MEMORY OF
THE LATE
ALFRED STURGES
WHO DONATED THIS
PARK
TO THE PEOPLE OF
OTAHUHU
1912

lookout here during early racial war scares in the 1840s. In the late 1850s, Ferdinand von Hochstetter named it after local landowner Robert Robertson. Sturges Park was named after Alfred Sturges, mayor of Ōtāhuhu, who donated the land to be a park in 1912.

Like many other Auckland volcanoes, Mt Robertson had a well and pumphouse used to help supply Ōtāhuhu with water until the early 1950s. The scoria cone's crater has been enlarged and the floor raised with fill to form the sports stadium of today. The crest of the scoria cone used to be 5 m higher on both the north and south sides but the top was taken off by quarrying in the 1950s and 1960s to produce a flat playing field next to the rugby clubhouse and the wide car park on the south side where concrete reservoirs once stood. The swampy moat has been drained and turned into playing fields used by Otahuhu College across Mangere Rd.

Volcanoes of
southern Auckland

⊕ Looking southeast over the two superb explosion craters with surrounding tuff rings of Pūkaki Lagoon (foreground) and Crater Hill (beyond). *Photo by Alastair Jamieson, 2018*

Most of the 16 volcanoes in this southern group erupted through the low-lying plains of the Manukau Valley during the Last Ice Age when the climate was cooler and sea level lower. Nine are primarily composed of an explosion crater and surrounding tuff ring, although three of these (Māngere Lagoon, Waitomokia and Crater Hill) also contained scoria cones that have been quarried away. Puhinui Craters comprises three small explosion craters and low tuff rings. The other seven volcanoes were composed of one or more scoria cones with associated lava flows. All the cones and most of the lava-flow fields have suffered from 20th-century quarrying, with complete loss of the Wiri, Matukutūreia, Maungataketake and Ōtuataua cones. Half of the cones on Puketūtū have gone, and there have been substantial bites taken out of tiny Pukeiti cone and the south's sentinel volcano, Māngere Mountain.

Some lava flows from Māngere, Ōtuataua and Matukutūreia are now in reserves (Ambury Regional Park, Ōtuataua Stonefields Historic Reserve and Matukuturua Stonefields Historic Reserve) that protect the only significant remnants of the once enormous stonefield gardening areas of pre-European Māori. Substantial portions of the remains of nine of these southern volcanoes are in public reserve, and much of Puketūtū will soon be opened. Hopefully it will not be too long before the one remaining volcanic jewel in the Auckland Volcanic Field that is still in private ownership – Crater Hill – will be purchased as a public park for the enjoyment of future generations.

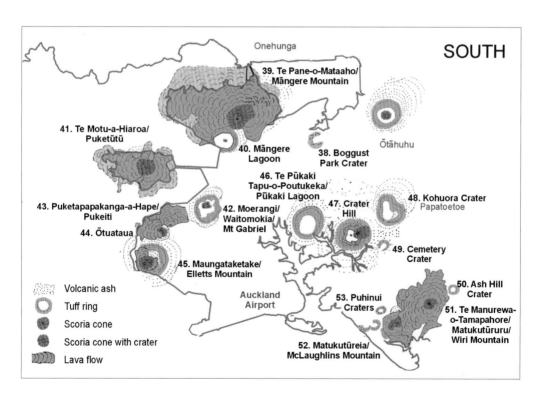

Boggust Park Crater

⊕ View west from above Harania Estuary looking into Boggust crater and tuff ring, 2018. *Photo by Alastair Jamieson*

Land status: The crater and part of the tuff ring crest are within Auckland Council's Boggust Park. The remainder of the tuff ring is beneath private properties.

What to do: Walk around the perimeter of the crater within the park; have a picnic or play an informal ball game.

Geology

This volcano was first recognised in 2011. It has a 300–400-metre-diameter crater surrounded on three sides by a 6–8-metre-high semi-circular tuff ring with steep inner slopes and gentler outer slopes. In places the crest of the tuff ring has been flattened off by bulldozing during subdivision in the 1990s. The tuff ring is breached to the northeast, presumably by the sea during the Last

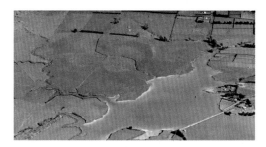

⊙ View southwest across Harania Estuary and Favona Rd bridge (right) in 1949 with Boggust Crater a shallow depression in farmland (centre). *Whites Aviation Collection, Alexander Turnbull Library*

Interglacial warm period, about 125,000 years ago, when the sea level was 5–6 m above the present. Today, the floor of the crater is about 5 m above high-tide level and slopes down to the edge of Harania Creek estuary on the Manukau Harbour. After eruption, Boggust Park Crater became a freshwater lake, before it was breached by the sea to become an intertidal lagoon for a few thousand years. Sea level dropped about 120,000 years ago and the crater became a swampy depression until 1 m of fill was added and drainage installed to make it a recreational sports field.

Human history

No Māori name for this feature is known. Boggust Park Crater has been named after the park in which it is located. The park is named after Ralph Boggust, former superintendent of Manukau Parks Department. In 1958, a major wastewater pipe from southeast Auckland to the new Māngere sewage treatment plant was buried across the middle of the explosion crater. From 1966 to the early 1990s, the crater was in the middle of the Māngere Hospital psychopaedic facilities, which had a number of buildings in the vicinity. Residential subdivision reached this area in the 1990s and at this time the bowl-shaped crater was set aside as a public reserve.

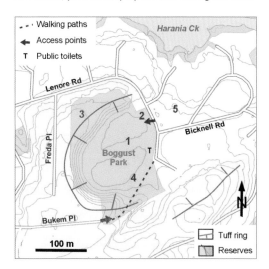

Places of interest around Boggust Park Crater:

1. Flat floor of naturally infilled crater.
2. Playground.
3. Flattened-off crest of northern part of tuff ring.
4. Basketball court.
5. Northeast part of tuff ring removed when breached by high sea level about 125,000 years ago.

Te Pane-o-Mataaho/
Māngere Mountain

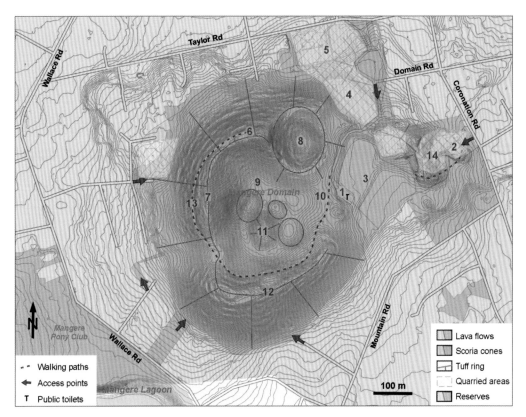

Lava flows
Scoria cones
Tuff ring
Quarried areas
Reserves

100 m

◉ Aerial view of Māngere Mountain scoria cone from the southeast in 2018 showing the extensive pre-European terracing on the crest and outer slopes. The smaller fountaining crater is on the upper right and the main crater in front to the left with the extruded lava plug in the centre.
Photo by Alastair Jamieson

Places of interest around Te Pane-o-Mataaho/Māngere Mountain:

1. Main Domain car park; clubhouse for Onehunga Mangere United Sports Club (soccer and softball).
2. Māngere Mountain Education Centre and Visitor Centre (open weekdays) in old Manukau City quarry. Also former brick water pumphouse. Scoria exposures.
3. Playing fields in re-landscaped area.
4. Children's playground, skateboard park and basketball hoop in former quarry.
5. Bridge Park Tennis Club and Bridge Park Bowling Club in former crater with scoria quarry to the east (car park extension).
6. Highest point on scoria cone and site of removed water reservoir.

7. Trig station on second-highest point.
8. Secondary fire-fountaining crater.
9. Main fire-fountaining crater surrounded by scoria ramparts.
10. Lowest part of scoria cone probably breached and mostly rafted away by lava flows.
11. Central tholoid (extruded plug) surrounded by three gas explosion craters.
12. High southern rim of scoria cone with extensive pre-European terracing and pits.
13. Shell midden (don't touch) and explanatory sculpture beside track up cone.
14. Partly quarried mounds of scoria that had been rafted here by lava flows from the west.

Land status: Virtually all of Māngere Mountain scoria cone lies within Māngere Domain, administered by Auckland Council. Most of the large lava-flow field is in residential housing, except the foreshore reserve along Kiwi Esplanade and Ambury Regional Park, both also administered by Auckland Council.

What to do: Take an energetic walk along the farm road to the summit; picnic; playground; playing fields; arrange a class visit and learning experiences (volcanic, archaeological, Māori culture, traditional and medicinal gardens) with Māngere Mountain Education Centre (www.mangeremountain.co.nz). See informative displays and artefacts at the Education Centre's Visitor Centre (free admission, 100 Coronation Rd).

Geology

Māngere Mountain is the largest and best-preserved volcano in the southern part of the Auckland Volcanic Field. It consists of a large, 105-metre-high scoria cone produced by fountaining from its large main crater about 50,000 years ago. A second, smaller fountaining crater can be seen on the northern rim of the scoria cone. A third, U-shaped breached crater and scoria ridge used to be present but was removed by quarrying in the 1950s–60s. During the latter phases of eruption a large quantity of fine scoriaceous ash was erupted from Māngere Mountain and the plume was blown to the northeast by strong southwest winds. This thick blanket of volcanic ash can be traced today mantling lava flows from One Tree Hill Volcano as far as 5 km away at Penrose.

Māngere Mountain is unique in Auckland with the presence of a small, 12-metre-high conical plug sitting over the vent in the centre of the main crater. After fountaining eruptions finished, the magma in the vent had time to cool and solidify into a near-solid basalt plug with still-fluid magma in the plumbing beneath. Gas bubbles rising up through the column of viscous magma gradually built up considerable pressure beneath the plug, sufficient to push the near-solid mass out of the volcano's throat, like a cork out of a bottle. As the plug was extruded, the pent-up gas escaped around its northern side creating three small circular gas-venting explosion craters. The escaping gas threw out many large globs of molten viscous lava that acquired aerodynamic spindle, spherical and corkscrew ribbon shapes as they cooled while hurtling through the air. Many

◑ The extruded plug and three surrounding small explosion craters in the middle of Māngere Mountain's main crater. These were the last features formed during the mountain's eruption.

⊙ The old brick pumphouse (operating 1932–1950s) still stands amidst the recreated traditional gardens in the grounds of Māngere Mountain Education Centre.

of these bombs, broken or whole, still lie scattered around the small craters and the inner walls of the main crater.

Towards the end of fountaining eruptions, the main crater was breached to the east by lava flows, which rafted away the scoria ramparts in the area of the present soccer fields and deposited them as rubbly mounds in the vicinity of the Māngere Mountain Education Centre and further afield. Large volumes of lava flowed from around the base of most of the cone, forming an extensive lava-flow field covering 500 hectares. It spread out as a wide apron of overlapping lava flows on

the gentle slopes of the forested Manukau River valley. These flows now form all the land beneath the suburb of Māngere Bridge, as far west as Ambury Regional Park. East of Māngere Mountain the lava flows extend over a kilometre beneath the route of the Southwestern Motorway.

Human history
The name Māngere originates from Ngā Hau Māngere – the 'lazy breezes' observed by Taikehu of the *Tainui* canoe when he landed below the mountain six centuries ago. The Māori name for the mountain is Te Pane-a-Mataaho, meaning 'the sacred head of Mataaho', god of volcanoes. Another name, Te Ara Pueru, specifically applies to the small quarried-away scoria ridge on the northeast side of the mountain. The scoria cone rim has two distinct high areas (along the south and northwest sides), both of which were extensively terraced and defended as separate pā. The eastern rim of the deep smaller crater was a communal food store in pre-European times with its two rows of rectangular storage pits.

An early European name for Māngere Mountain was Mt Elliott but this is no longer in use. Despite its present appearance, Māngere Mountain scoria cone did not escape some quarrying in European times. Between 1900 and 1964, six separate scoria pits operated at different

⊙ In a depression at the northeast base of Māngere Mountain is a rehabilitated quarry pit used as overflow parking for Bridge Park tennis and bowling clubs (on right). These club grounds were the floor of a U-shaped breached crater. The eastern scoria ramparts of this crater were removed by quarrying in the 1950s–60s.

times around the cone's lower northern and
eastern slopes. The largest removed a prominent
northeastern ridge, which is now the site of the
Bridge Park bowling and tennis clubs. Other
quarries removed scoria mounds and knolls
either side of the entrance driveway to Māngere
Domain (now playground and soccer field) and
back of the Manukau City Works Depot (now
the Māngere Mountain Education Centre).
Extending across the Manukau lowlands, the
Māngere Riding of Manukau County ran its own
independent water supply using a borehole at
Māngere Mountain from 1932 to the 1950s.

◗ A large volcanic bomb in Māngere Mountain's main crater. Numerous bombs were ejected from the three explosion
craters as the plug was forced out of the volcano's throat.
◗ View northeast across the smaller fountaining crater in 1960 showing the extensive quarrying that removed a whole
scoria ridge and the breached northeastern fountaining crater, now occupied by Bridge Park tennis and bowling clubs.
Whites Aviation Collection, Alexander Turnbull Library

Kiwi Esplanade pahoehoe flows

Land status: Publicly accessible coastal strip and intertidal shore, best seen opposite
34–59 Kiwi Esplanade.

◒ The ropey pahoehoe surface on lava-flow lobes from Māngere Mountain are seen in many places alongside Kiwi Esplanade, Māngere Bridge.
◓ Meandering lava-flow lobes from Māngere Mountain are a common sight along the foreshore of Kiwi Esplanade.

Most of the lava from Māngere Mountain was hot and fluid and would have flowed rapidly off downslope at speeds of up to 10 km/h. The outside of these red-hot flows chilled quickly, forming a black, smooth or ropey skin that wrinkled up into small curved ropey ridges, called pahoehoe. The best examples in Auckland of pahoehoe ropey lava surfaces and numerous branching lava-flow lobes can be seen adjacent to Kiwi Esplanade, Māngere Bridge. Since forming, the surfaces of these typical pahoehoe flows were protected from weathering and erosion by a thick covering of volcanic ash that has only recently been eroded by the sea to expose them to view.

Ambury Regional Park lava flows

Land status: Ambury Regional Park (Ambury Farm) is administered by Auckland Council.

What to do: A wonderful place for a walk or bike ride in the coastal countryside within the city; picnics; barbecues; explore the lava-flow features; bring binoculars and watch the shorebirds; give the family a taste of farm animals; campground; children's playground (Purata Park).

Geology

Ambury Regional Park (also known as Ambury Farm) occupies a large part of the western ash-covered lava-flow fields of Māngere Mountain, which spewed out from the base of the volcano ~50,000 years ago. For most of the time since then they would have been covered in native forest. In the farm paddocks and around the coastal fringe there are numerous exposures of basaltic lava flow. In some places there are low arches with entrances to small lava caves that formed as hot fluid lava flowed out from beneath a cooled solid roof of rock. Around the foreshore one can see examples of ropey pahoehoe surfaces, branching lava lobes and in one place the moulded impression of a large tree trunk that was caught up and rafted along by a flow.

Human history

The forest was cleared by Māori colonisers within the last 700 years. In pre-European times the rich soils were extensively cultivated with stone mounds and rows clearly visible, especially in the northeastern part of the park near Kiekie Rd. Around the coastal fringe there is archaeological evidence of numerous pre-European fishing settlements.

⊕ Aerial view eastwards across Māngere Mountain lava-flow field where it underlies Ambury Regional Park (foreground) and the suburb of Māngere Bridge (left). Note the rocky foreshore formed from the lava-flow toes. *Photo by Alastair Jamieson, 2018*

◔ The low entrance to one of a number of small lava caves within the flows from Māngere Mountain (background) that underlie Ambury Regional Park.

The name Ambury Park comes from Ambury Milk Company, which operated the property as a town milk supply farm from 1893 to 1965. The land was purchased by the Auckland Regional Authority (1965–73) as a buffer to the adjoining Māngere Wastewater Treatment Plant oxidation ponds.

Bird life

Ambury Park is popular with birdwatchers, primarily for the migrant wading birds that roost in their thousands around its coastal fringes and on the shell islands just offshore when their feeding grounds are immersed at high tide. The Arctic waders, primarily eastern bar-tailed godwit and lesser knot, but also turnstone, golden plover, curlew sandpiper, long-billed curlew and whimbrel, are present in significant numbers between October and March. In autumn and winter, South Island migrant waders are common, particularly pied oystercatcher, wrybill and pied stilt. Amongst the nearly 90 species of birds recorded from the park are the spectacular royal spoonbill and cattle egret. Also commonly seen around the rocky coast and small shelly beaches are pied, black, little black and little shags and white-faced herons. Feeding in the paddocks you are likely to see pūkeko, spur-winged plover, swallow, ducks, finches, myna, starling and skylark.

◑ This 3-metre-long smooth surface within a basaltic lava flow on the foreshore of Ambury Park is the moulded impression of the trunk of a tree captured in the cooling lava as it was rafted along on the top of the flow.
◔ Unusual pahoehoe lava-flow surfaces in the foreshore of Ambury Park may have been produced by differing flow rates within one wide flow.

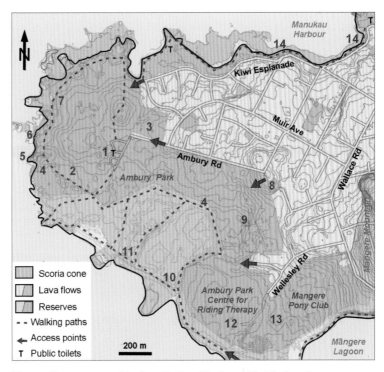

Places of interest around Ambury Regional Park and Kiwi Esplanade:

1. Car parks, information boards, farm barn, ranger office.
2. Ambury campground.
3. Bull Paddock Lava Cave. Available for group visits (ask the ranger).
4. Small lava caves in paddocks, usually too low and small for access.
5. Fallen tree trunk impression in lava flow, visible at low tide.
6. Ropey pahoehoe surface on lava flow; picnic shelter and bird hide.
7. Informal (often windswept) foreshore walk around the coast, no bikes (follow yellow markers).
8. Purata Park children's playground.
9. Pre-European stonefield gardens with stone heaps (follow markers).
10. Watercare Coastal Walkway to Māngere Lagoon explosion crater and Ōtuataua Stonefields.
11. Bird hide.
12. Ambury Park Centre for Riding Therapy on Māngere Mountain lava-flow field.
13. Mangere Pony Club on Māngere Mountain lava flows.
14. Good examples of pahoehoe lava visible at low tide, opposite 32–45 and 55–64 Kiwi Esplanade.

Māngere Lagoon

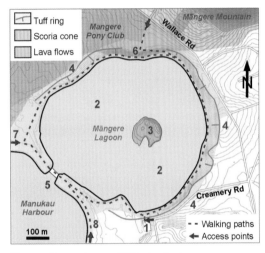

● View of the restored Mängere Lagoon explosion crater from the south with Mängere Mountain scoria cone beyond. *Photo by Alastair Jamieson, 2018*

Places of interest around Mängere Lagoon:

1. Car park.
2. Intertidal flats of the explosion crater floor. Beware deep mud.
3. Restored small, low-lying scoria cone.
4. Tuff ring around north and east sides of lagoon.
5. Breach in tuff ring now route of western interceptor sewer line.
6. Lava flow from Mängere Mountain can be seen where it spilled over the tuff ring crest and into the basin; beside track to Wallace Rd.
7. Watercare Coastal Walkway to Ambury Park.
8. Watercare Coastal Walkway towards Ōtuataua Stonefields.

Land status: Māngere Lagoon and tuff ring is publicly owned reserve administered by Watercare.

What to do: Walk around the perimeter track inside the crater (2 km) or continue on along the Watercare Coastal Walkway in either direction.

Geology

Māngere Lagoon is a castle-and-moat volcano, consisting of a small central scoria cone located in the centre of an explosion crater and surrounding tuff ring. Māngere Lagoon erupted not long before Māngere Mountain about 50,000 years ago. Lava from Māngere Mountain filled up the valley between the mountain's scoria cone and the lagoon's tuff ring and some of the lava then spilled down the inside of the tuff ring into the lagoon's crater.

The tuff ring was breached by rising sea level about 7500 years ago. The crater floor would have rapidly filled with marine mud brought in suspension with the tides, and by the time of human arrival its 36 hectares were fully intertidal and fringed by salt marsh, and partly colonised by mangroves.

Human history

European farmers let their stock graze the tidal flats and erected a fence right across them, allowing the stock to graze on the cone as well. In the late 1950s, Auckland City began construction of a new sewage disposal scheme to be located at Motukorea/Browns Island in the Waitematā Harbour. A campaign by mayoral candidate Sir Dove-Myer Robinson saved this beautiful island volcano and the Waitematā Harbour from a terrible fate. In its place, 500 hectares of oxidation and sludge ponds were constructed on the less-favoured, and at the time less-populated, Māngere foreshore of the Manukau Harbour. Sacrificed were 13 km of coastline and two of the nearby volcanoes, which were extensively damaged during construction. The small scoria cone in Māngere Lagoon was decapitated and became

the hub of a radiating network of embankment spokes constructed to subdivide the crater floor into sludge treatment ponds.

For nearly 40 years, Māngere Lagoon played its part in treating and disposing of Auckland, Waitakere and Manukau cities' sewage. The need to increase capacity and upgrade the plant led planners in the 1990s to propose a switch to the use of new land-based technology. As a result restoration of the oxidation and sludge ponds to their former tidal flats occurred between 1998 and 2005 – the largest coastal rehabilitation project in New Zealand history.

Today, Māngere Lagoon explosion crater is once again a tidal lagoon with the cone restored to approximately its former shape. The only remaining unnatural modifications are the major sewerage pipeline running along the seaward side of the lagoon, with incoming and outgoing tidal waters siphoned beneath it, and the presence of additional flat land built up above high tide and attached to the scoria cone island as an artificial high-tidal roost for wading birds, which is little used and should be removed.

◑ View from the east over the small scoria cone in Māngere Lagoon in 1955, prior to its demolition for use in the sewage treatment operations. Note the fountaining crater surrounded by low scoria rim with two pre-European kūmara pits on the highest part. *Whites Aviation Collection, Alexander Turnbull Library*
◑ Māngere Lagoon in 1983 during its time as six sludge ponds – part of Auckland's sewage treatment system. The central hub is the decapitated scoria cone in the middle of the explosion crater. *Photo by Les Kermode*
◑ Restoration of Māngere Lagoon's sludge ponds in progress in the early 2000s, viewed from Māngere Mountain.

Te Motu-a-Hiaroa/
Puketūtū

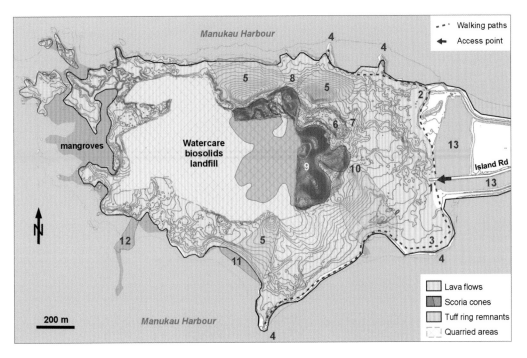

Places of interest around Te Motu-a-Hiaroa/Puketūtū Island:

1. Car park and start of walking tracks.
2. Northern coastal track.
3. Southern coastal track.
4. Remnants of bunds that separated oxidation ponds which covered the intertidal area between the island and the mainland from late 1950s to early 2000s. The area has now been restored.
5. Smooth slope is tuff ring remnant* (visible from end of northern track).
6. Kelliher Homestead now used as reception venue.*
7. Historic basalt stables.*
8. Additional small houses and stables.*
9. Remaining scoria cones* (visible from walking tracks and from a distance).

10. Disused historic basalt-walled water tank.*
11. Mid-tidal exposures of tuff containing baked blocks of sandstone that were thrown out of the erupting vent (accessible by walking around mangroves in gumboots at low tide)
12. Lava flow intruding beneath and through up-domed tuff (accessible in gumboots at low tide)
13. Holding ponds where treated wastewater from Māngere Wastewater Treatment Plant is held to be released into Manukau Harbour with the outgoing tide.

*Not publicly accessible in 2019.

◑ Te Motu-a-Hiaroa/Puketūtū Volcano is the only island volcano in the Manukau Harbour. View from the northwest in 2018. Note the quarry is being filled with solid waste from Watercare's sewage treatment plant.
Photo by Alastair Jamieson

◔ This reef on the southern shore of Te Motu-a-Hiaroa/Puketūtū Island is a narrow lava-flow ribbon that initially pushed its way under soft sediment and tuff before breaking out at the surface in the foreground.

Land status: The whole island is owned by Te Motu a Hiaroa Charitable Trust. In 2019, parts are leased to Watercare for biosolids disposal; Living Earth for composting operations; a reception venue; and Auckland Council to prepare the remaining parts of the island for opening as a public park.

What to do: Walk or cycle the coastal paths; take your binoculars and watch the seabirds feeding on the intertidal flats. There will be much more to do and see once the public park is opened.

Geology

Te Motu-a-Hiaroa/Puketūtū Volcano is the only island volcano in the Manukau Harbour. When it erupted about 30,000 years ago, however, there was no Manukau Harbour as the climate was much colder and sea level considerably lower. Like many other Auckland volcanoes, Puketūtū began with wet explosive eruptions that threw up an 800-metre-wide tuff ring around its crater.

Subsequent eruptions completely filled this crater and only four short sections of tuff ring arc remain, forming smooth, gently sloping triangular hills on the island's north and south sides.

The main eruptions at Puketūtū were fire-fountaining from a number of closely spaced vents that produced a central complex of scoria cones and mounds to a maximum height of 65 m. Some of the youngest eruptions built the

⊙ A thin dike of solidified basaltic lava intrudes through brown volcanic tuff on the southern shore of Puketūtū Island. The dike is a small sheet of lava that was squeezed up along a vertical crack above a narrow lava flow that pushed its way under soft sediment and ash as it flowed away from the vent.

⊙ A block of Waitematā Sandstone in a bed of tuff on the southwest side of Puketūtū Island. It was ripped from the wall of the vent during an explosive eruption. The heat of the eruption cloud baked the outer zone.

easternmost cone, which still has a small summit crater. Towards the end of these fountaining eruptions, large volumes of gas-poor lava flowed out from a number of vents between and around the lower parts of the scoria cones, building up a thick apron of overlapping lava flows that completely surrounded the central cones, bursting through and overflowing the tuff ring in many places.

Intertidal exposures of rock along the southern coast of the island provide insights into some of the volcanic processes that produced Puketūtū. Exposures of bedded tuff at high-tide level contain many projectile blocks of Waitematā Sandstone that were ripped from the walls of the volcano's throat during wet explosive eruptions. Some sit in the original impact depressions that they made in the soft ash when they landed. Halfway along the south shore is a narrow ribbon-like lava flow that has pushed its way into the soft sediment beneath a thin layer of volcanic ash. This intrusion heated up and baked the sediment in close contact with the flow, creating columnar-jointed, orange-coloured natural brick around it. As this flow pushed its way along, the sediment and tuff layers above it were domed upwards, forming an anticline that was fractured along its crest. Some of the lava squeezed up these fractures and cooled to form narrow sections of basalt dike.

Human history

Puketūtū played a significant role in the history of the *Tainui* canoe's travels through the Auckland area about 600 years ago. It was here that the canoe was hauled ashore to be repaired. The Māori name for the island is Te Motu-a-Hiaroa, meaning 'the island of Hiaroa', the wife of Hoturoa, the commander of the *Tainui* canoe. Puketūtū means 'hill of the tutu scrub' and refers to the summit of a prominent triangular scoria peak, which has been quarried away. The largest cone remaining was named Te Taumata-o-Rakataura – 'the summit of Rakataura' – because Rakataura, the leading tohunga on the *Tainui*, carried out ritual ceremonies on the summit to placate the spirits of the land and to ensure the safety of his people.

The island's disused early European name was Weekes Island after Henry Weekes, the first European owner (1848–49). John Logan Campbell also owned the island for some years (1852–95). The most recent owners have been the Kelliher family, initially Sir Henry (1938–91) and since his death the Kelliher Charitable Trust. European owners farmed the island, building extensive drystone basalt walls from the abundant rocks scattered over the surface of the ancient lava flows. Sir Henry's period of ownership is best remembered for the premier horse stud he ran on the island. Up until 1958, Puketūtū Volcano was largely intact with only a small quarry on the northeastern side of the cones near the water pump. In that year the late Ernie Searle wrote: 'May Puketutu long continue to be one of Auckland's beauty spots.' But in his 1964 book *City of Volcanoes*, he added the following sad note:

[D]espoliation of this lovely island is already under way. In the last few weeks excavations of large quantities of scoria, for use on the runways of Mangere Airport, at present under construction, have bitten deeply into the heart of the island. One small cone has already gone completely and the dominating peak, on which was sited the ancient Maori fortification of Puketutu, has been destroyed and soon will be levelled to its base. No doubt attempts will be made to heal the scars and cover the ugly wounds but the effects of man's rude surgery cannot be obliterated. It is perhaps inevitable that twentieth century man should value a flat strip of concrete more than the charming but 'useless' hills with which nature so lavishly adorned the city.

Quarrying continued unabated until 2008. More than half the scoria cone complex disappeared and most of the lava flows on the western and southern parts of the island were removed – some replaced by cleanfill from around the city. The easternmost scoria cones and lava flows that underlie the horse paddocks on the northeastern lava flows remain, although all the surface rock was long ago collected up and used to construct drystone basalt walls.

In 2010, Watercare was granted resource consent to use the old quarries as a repository for the solid wastes from the adjacent sewage treatment plant. At the same time ownership was transferred to the Te Motu a Hiaroa Charitable Trust who plan to build a marae on a small part of the northern side and lease the currently unused parts to Auckland Council as a public park. In several decades' time, when Watercare has finished using the old quarries for fill, there will be a new 30-metre-high landfill plateau surmounted by small hills that will hopefully restore as accurately as possible the shapes of the scoria cones and mounds that were quarried away.

◔ Watercare's biosolids landfill operation viewed from the top of Te Motu-a-Hiaroa scoria cones in 2018.
◑ View westwards across the northern side of Te Motu-a-Hiaroa/Puketūtū Island prior to any quarrying, showing the smooth tuff ring slope (right), scoria cones (left) and lava flows (beyond). *Geoff Fairfield, 1938, University of Auckland*
◔ Te Motu-a-Hiaroa/Puketūtū Island from the west in 1946 before quarrying had removed any scoria cones or lava flows and prior to the planting of stands of exotic pine trees. *Whites Aviation Collection, Alexander Turnbull Library*

Moerangi/Waitomokia/
Mt Gabriel

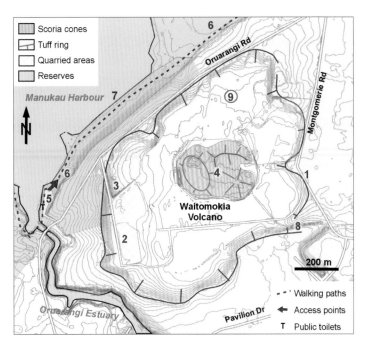

Places of interest around Waitomokia:

1. Vehicle entrance off Montgomerie Rd to Villa Maria Vineyard Café.
2. Villa Maria Vineyard Café and car park (accessible when open) inside explosion crater.
3. Road cuttings through tuff ring.
4. Site of three former small scoria cones (private industrial yard).
5. Oruarangi Creek mouth car park.
6. Watercare Coastal Walkway.
7. High-tidal exposures of tuff with projectile blocks of sandstone and conglomerate.
8. Small overflow stream eroded through tuff ring to Oruarangi Estuary.
9. Site of small steam explosion crater, now destroyed.

❻ Waitomokia explosion crater and low surrounding tuff ring from the south.
Photo by Alastair Jamieson, 2018

◔ Aerial view from
the north of the three
small welded scoria and
spatter cones (soon after
quarrying had started)
in the centre of the large
explosion crater. Note the
two explosion pits on the
tuff ring in the foreground.
*Jack Golson, 1957, University
of Auckland*

Land status: Almost all the crater and tuff ring crest is privately owned, although public access is available to the Villa Maria Vineyard Café inside the crater (118 Montgomerie Rd). The western outer slopes of the tuff ring (seaward of Oruarangi Rd) are publicly accessible on walkway.

What to do: Examine the tuff and projectile blocks along the foreshore from the car park. Walk or cycle the coastal walkway; take your binoculars and watch the birds; do a wine tasting or have coffee at Villa Maria.

Geology

Waitomokia was originally a fine example of a castle-and-moat volcano. The volcano consists of an elliptical explosion crater (700 x 1000 m across) surrounded by a 15–25-metre-high tuff ring with three small cones (castle) in the centre. The elongate crater was probably produced by two or more vents that erupted along a north–northeast-trending fault. The irregular scalloped outline of the crater was formed by the curved head scarps of sections of tuff ring that had slumped back into the deep hole during or not long after the eruptions. The northeastern crest of the tuff ring was pockmarked by a line of seven or eight small (10–25-metre-diameter)

steam-escape craters that presumably erupted late in the volcano's history.

The three small cones, each rising to just over 30 m, were produced by fiery explosive eruptions from three vents in the centre of the explosion crater. The two eastern cones were conical, whereas the southwestern one was a perfectly formed steep-sided welded scoria and spatter cone with a funnel-shaped, 18-metre-deep crater inside its narrow-rimmed crest. Geophysical studies have found that dense basalt partly fills the explosion crater and seems to extend beneath the central cones. Presumably this basalt is a solidified lava lake (ponded lava flows) that exuded up into the crater with associated fiery explosive eruptions above the vents, throwing out ragged scoriaceous lumps of magma that built the cones on top of the lava lake's solidified crust.

Following the cessation of eruptions, the large crater and its solidified lava lake partly filled with water to become a freshwater swamp. Over thousands of years, decaying vegetation and sediment buried all trace of the underlying basaltic lava. Volcanic ash from Waitomokia has been identified in Pūkaki Lagoon and dates the eruption at close to 20,300 years ago.

Human history
The name Waitomokia means 'water seeping into the ground' and refers to the ponds in the crater that commonly dried up in summer. The small cones in the centre were known as Moerangi. Some long-time locals still know this volcano by its early European name of Mt Gabriel, after an early settler. Pre-European Māori recognised the natural defensive advantages of building forts on the two conical cones in the swamp, which were extensively terraced with kūmara storage pits dug on the crest of the cratered cone.

The scoria cones (castle) were removed by quarrying in the late 1950s for use in the creation of the Māngere sewage treatment works across the road. Quarrying, road works and development around the crest of the tuff ring have removed most of its original detail. Villa Maria has established a major vineyard and winery in the southern half of the crater and this is sometimes used for public concerts on summer evenings.

🔾 Projectile block of cream and grey Waitematā Sandstone within tuff on the Manukau Harbour foreshore of Waitomokia Volcano. The block was ripped from the wall of the volcano's throat and blasted out of the crater by an explosive eruption, and broke into pieces when it landed 500 m away.
🔾 Sketch by Ferdinand von Hochstetter in 1859 of the view from the northeast of Waitomokia crater with its three small central cones surrounded by swamp.
Courtesy of Hochstetter Private Collection, Basel

Puketapapakanga-
a-Hape/Pukeiti

⊘ Pukeiti Volcano fills most of the photo with lava flows fanning out from the small spatter cone-capped vent that has been quarried around its sides. *Photo by Alastair Jamieson*

Land status: The small cone and lava-flow field are almost entirely within Ōtuataua Stonefields Historic Reserve, which is managed by Auckland Council as a farm park.

What to do: Walk and explore the remains of two volcanic cones, their lava flows and the pre-European stonefield gardens on top of them.

Geology
Pukeiti Volcano comprises a 1-kilometre-wide, 800-metre-long fan of lava flows that spread northwards from a small vent, over which a small, 10-metre-high scoriaceous spatter cone was built. This cone is the smallest remaining in the Auckland Volcanic Field and is different in composition from the larger scoria cones. Small quarry faces in its sides show that it is made of rather dense basalt spatter that fountained out of the vent as voluminous lava poured out from around its northern base. A small dish-shaped crater still remains on the summit. Pukeiti's lava flows are mantled in volcanic ash, unlike those of its neighbour Ōtuataua. This suggests that Pukeiti is the older of the two volcanoes that possibly both erupted about 15,000 years ago.

⊙ View northwards from Ōtuataua scoria cone in 1899 showing the small spatter cone and crater of Pukeiti among stone-walled fields. *Hugh Boscawen, Auckland Museum*
⊙ Example of the fused scoriaceous spatter that Pukeiti cone is made of.

On the surface of the ash-covered lava flows there are several trenches that run downhill away from the small cone. These record the route of individual flows, where the crusted-over roof collapsed into the lava cave tube after the molten magma had drained out. Also within the flows there is a 90-metre-long lava cave with its roof still intact. It is known as Lino Lava Cave, as cavers placed a strip of lino on the floor to protect their knees from the sharp basalt as they crawled through a low section of tunnel. Today, entry into this cave is prevented by a metal grille across the entrance.

Human history
This small cone has several Māori names. Pukeiti means 'small hill', whereas its more sacred name is Te Puketapapakanga-a-Hape, which refers to 'the flat-topped hill where Hape arrived'. Hape was another name for Rakataura, the leading tohunga on the *Tainui* canoe when it visited Auckland. Several elongate pits around the flanks of the small cone were made by early European quarrying.

Places of interest around Pukeiti:
See page 273.

⬆ The small Pukeiti spatter cone in Ōtuataua Stonefields Historic Reserve.
⬇ This trench with levee sides is the route of a lava flow from Pukeiti where the solid roof of the internal lava tube collapsed and was rafted away by the flowing molten lava.

Ōtuataua

⬆ The lava-flow field of Ōtuataua Volcano stretches westwards to the shores of the Manukau Harbour. Beyond to the right is the quarried-out, cup-shaped remains of the scoria cone that was formed above the vent. Stone heaps, rows, walls and enclosures constructed by pre-European Māori and early European farmers cover the lava flows. *Photo by Alastair Jamieson, 2009*

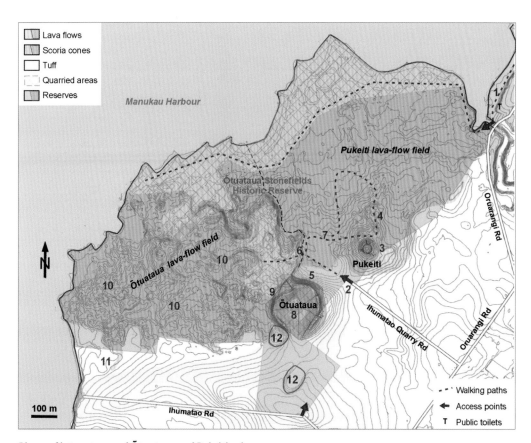

Places of interest around Ōtuataua and Pukeiti volcanoes:

1. Oruarangi Creek mouth car park, Watercare Coastal Walkway.
2. Ihumatao Quarry Rd main entrance to Ōtuataua Stonefields Historic Reserve.
3. Pukeiti spatter cone, partly quarried. Please keep out of the crater as it is a sacred site.
4. Surface trench with levee sides is the route of a Pukeiti lava flow where the roof of the internal lava tube has collapsed and been rafted away by the flowing lava.
5. Avocado orchard.
6. Signpost for start of three self-guided walks.
7. Self-guided geology walk on Pukeiti Volcano.
8. Quarried-out inside of Ōtuataua scoria cone; views from crest.
9. Face of old scoria quarry with thin intrusive basalt dike and three thin lava flows within the scoria cone sequence.
10. Rocky surface of Ōtuataua lava flows with numerous pre-European stone heaps, rows and house sites. Also European drystone basalt walls.
11. Narrow lava-flow lobe from Ōtuataua lava field.
12. Shallow explosion craters.

For maps and details about the history, geology and botany of Ōtuataua Stonefields Historic Reserve, see www.bit.ly/2Xce4sc

◉ View from the southwest showing Ōtuataua scoria cone with its top half removed by quarrying, 1964. Unquarried, as yet, and clearly visible are two lava flows that flowed out from the breached crater on the left side of the cone. The flows are separated by a levee ridge covered with rectangular kūmara pits. *Whites Aviation Collection, Alexander Turnbull Library*

Land status: The scoria cone and lava-flow fields all lie within Ōtuataua Stonefields Historic Reserve, which is managed by Auckland Council as a farm park.

What to do: Walk and explore the remains of the quarried-out scoria cone, the lava-flow field and pre-European stonefield gardens over them.

◉ Three thin lava flows that spilled out from a narrow intrusive dike feeder from part-way up the side of Ōtuataua scoria cone. These can be seen in an old quarry cutting in the scoria cone sequence on the southwest side of the cone.

Geology

Remnants of a small explosion crater and tuff ring, formed at the start of eruptions, are present to the south and east of the scoria cone. In winter a temporary pond or swampy wetland develops in the remains of this crater.

The main phase of eruptions about 15,000 years ago entailed fountaining from the single vent, which produced a 64-metre-high scoria cone. Lava poured from around the western foot of the growing cone and breached the small summit crater, creating a horseshoe-shaped cone. The lava spread in a narrow, 800-metre-long fan to the west and northwest, and today its toe forms the Manukau Harbour foreshore of the Ōtuataua Stonefields Historic Reserve.

◑ Stone heaps and rows in Ōtuataua Historic Stonefields Reserve that were created in pre-European times, when Māori used the rich volcanic soil among the rocky lava-flow outcrops for gardening. Freestone walls were probably built by early European farmers.

The surface of these flows is rough, irregular and rocky. In places there are trenches, lava-flow rolls and small crevasses where congealed lava was left behind as more molten material flowed on.

Human history

The former European name for Ōtuataua Volcano in the 1930s was, unfortunately, Quarry Hill. Most of Ōtuataua's scoria cone was quarried away in the 20th century and two quarries in the 1950s–70s removed some of the lava-flow basalt in the west. In the 1950s, some of the scoria was used in building the nearby Auckland Airport runway. Basalt from the toe of the lava flows on the foreshore was used to build some of the embankments between the former oxidation ponds in the adjacent Manukau Harbour in the 1950s. In the 1960s, basalt from the upper lava flows was taken to help build the causeway for the Northwestern Motorway between Waterview and Te Atatū.

In 1999, the rocky Ōtuataua and Pukeiti lava-flow fields were purchased from their long-time European farming families to become a historic reserve. The last of the scoria was taken out of the old Ōtuataua scoria cone quarry at this time and its newly smoothed shape was planted in grass and, unfortunately, resembles a large crater. Hopefully this hole will be filled with cleanfill and the original pre-European terraced shape of the cone and pā will be restored to become the historic reserve's centrepiece.

The 100-hectare Ōtuataua Stonefields Historic Reserve is one of two remaining areas of lava-flow fields that were extensively modified and used for gardening in pre-European times. Originally, over 8000 hectares of stonefield gardens were present in the Auckland Volcanic Field, but now less than 0.2 per cent remains. The archaeological features in the Ōtuataua Stonefields are remnants of a complex land system that stretched from the volcanic cone to the sea and reveal a sequence of hundreds of years of occupation prior to the arrival of Europeans.

Maungataketake/
Elletts Mountain

This deep quarry pit is all that remained of Maungataketake's scoria cones in 2018. Its lava-flow field stretches out beneath the paddocks in the foreground and to the right and its low tuff ring arcs around beyond the quarry. *Photo by Alastair Jamieson*

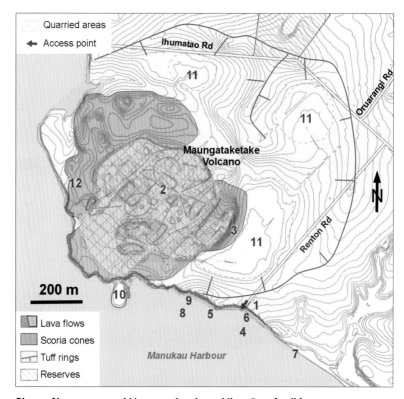

Places of interest around Maungataketake and Ihumātao fossil forests:

1. Steps down to foreshore and fossil forests at end of Renton Rd.
2. Quarried-away site of Maungataketake scoria cones.
3. Unquarried section of lava-flow ridge.
4. Fossilised in situ kauri tree stumps and fallen trunks in foreshore. The overlying peat that buried them has been eroded away by the sea.
5. In situ stump of kauri tree buried by black peat and overlain by tuff in base of cliff.
6. In situ tree trunk that has been knocked sideways, decapitated and buried in tuff by base surge eruptions; just left of the foot of the steps.
7. Numerous hollow moulds of branches and small upright tree stumps in tuff in low cliffs.
8. Blocks of hard limestone on the sand

flats. These were ripped from the walls of the volcano's throat and blasted out in explosive eruptions. Until recently they were buried in tuff that has been eroded from around them by the sea.

9. White pumiceous ash from eruptions in the centre of the North Island several hundred thousand years ago seen in the base of the cliffs. The older kauri forest grew in this material.
10. Artificial flat-topped prominence made of basalt blocks in the 1950s as a new high-tide roost for seabirds to attract them away from the Auckland Airport runway.
11. Sites of shallow explosion craters within the main tuff ring. Their flat floors are now filled with swamp sediment.
12. Site of Wesleyan Mission Station, 1847–63.

Land status: The entire footprint of
Maungataketake Volcano is privately owned.

Geology

Maungataketake erupted in all three styles about
90,000 years ago, according to Ar-Ar dating.
The initial wet explosive eruptions produced an
800-metre-diameter crater surrounded by a low
tuff ring. At the base of the tuff sequence are the
remains of forest that was killed and buried by
the early phases of Maungataketake's eruption.
Some tuff layers are full of small accretionary
lapilli or volcanic hailstones (1–2 mm in
diameter). In other places there are blocks of
country rock (the background rock through
which the volcanoes erupted) that have been
ripped from the walls of the volcano's throat and
blasted out of the vent. These blocks give us an
insight into the underlying geology and include
numerous examples of Waitematā Sandstone
and associated grit (Parnell Grit). Unique to

↑ A 30 cm block of algal limestone (Te Kuiti Group,
Oligocene age, ~30 million years old) that was ripped from
the walls of Maungataketake Volcano's throat and thrown
high in the air, landing in the soft surrounding ash of the
tuff ring. A number of hard limestone blocks like this have
been eroded out of the tuff cliffs and can now be seen on
the foreshore.
↓ The profile of Maungataketake cones viewed from the
east in 1949, prior to the start of quarrying. *Whites Aviation
Collection, Alexander Turnbull Library*

Maungataketake scoria cone complex from the southeast in the 1940s, prior to the start of quarrying that entirely removed it. *Whites Aviation, Auckland Museum*

Maungataketake is the presence of a number of large blocks, up to a metre across, of crystalline limestone that now sit out on the sand flats because they are so resistant to erosion. This is the only place where limestone (Te Kuiti Group, Oligocene age, ~30 million years old) is known to occur at depth beneath Auckland City. The sites of three shallow, largely infilled explosion craters, each 200–300 m in diameter, are still discernible within the main crater.

Magma continued to rise after all the groundwater was used up, and dry fountaining of frothy scoria followed. This built an unusually shaped scoria complex with two peaks, up to 70 m high. How many vents were involved is unknown, but two craters on the western side recorded the site of the last ones. Lava also welled out from the base of the cones, forming a thick flow that did not escape the explosion crater and encircled the scoria cones on all but the north side.

Human history

Maungataketake is Māori for 'everlasting mountain', which unfortunately it is not, as all that is left of the scoria cone today is a giant hole in the ground. The cones were extensively terraced by pre-European Māori and the highest peak was more heavily defended as a fortified pā.

The first Europeans settled in this part of Auckland in the mid-1840s and established the Wesleyan (Methodist) Mission at Ihumātao on the shores of the Manukau Harbour, west of Maungataketake. The European name for this volcano comes from the long-time owners, the Ellett family.

As the supply of scoria to the growing city began to run out from closer sources, quarrying proper began at Maungataketake in 1962. By the 1990s, the scoria cones were completely flattened, but quarrying continued below ground level, creating a large hole as the lower portions of the cones within the explosion crater were removed. The land is zoned to become an industrial subdivision.

Ihumātao Fossil Forest

Land status: The fossil forests are located in the publicly accessible intertidal foreshore and low cliffs of the Manukau Harbour, down steps from the end of Renton Rd, off Ihumatao Rd, Māngere.

In the foreshore and low cliffs at the end of Renton Rd are the fossilised remains of two ancient forests. Both grew during slightly colder times of the ice ages when sea level was lower than at present and the Manukau Harbour only existed as a broad, forested valley draining out past Whatipu to the coast.

Evidence of the older forest consists of numerous large kauri tree stumps, up to 2 m in diameter, with their roots still anchored in buried soil. All around them are the fallen trunks of these forest giants, some up to 30 m long, straight and branchless, preserved within the black lignite of an ancient peat swamp. The presence of golden chunks of kauri gum throughout the lignite and in cracks within some stumps confirms that these are the remains of a once superb kauri forest that grew in the Manukau Valley, 100,000–200,000 years ago. The forest probably died because the drainage of the area changed and a shallow swamp formed and invaded the kauri stand, killing the trees. The dead trees probably survived for many years before, one by one, they succumbed in strong gales and crashed to the ground to become waterlogged and preserved in the surrounding peaty swamp. The remains of 98 different species of beetle have been found in the upper part of the peat swamp deposit.

The second, younger fossil forest grew on the same site as the first, after the peat swamp had filled and become dry land. This forest was killed, buried and preserved by base surges and volcanic ash that erupted from nearby Maungataketake Volcano about 90,000 years ago. Its remains can be seen in the low sea cliffs. This was a mixed podocarp and broadleaf forest with numerous

⊙ This tree was partly pushed over and lost its top and branches during early base surge pyroclastic flows of volcanic ash and gas from nearby Maungataketake Volcano. It was subsequently buried by layers of volcanic ash. In recent decades coastal erosion of the low cliffs of tuff has uncovered this in situ tree trunk near the steps down to the beach at the end of Renton Rd, Ihumātao.

⊙ Rimu leaves like these are common in the lower layers of tuff at Ihumātao. Forest dominated by rimu was growing here on top of the dry peat swamp when Maungataketake Volcano erupted 90,000 years ago. The initial base surge eruptions stripped the leaves from the branches and they fell to the ground and were buried and fossilised in the lower layers of volcanic tuff. Photo 10 cm wide.

⊙ This fallen trunk of a large kauri tree is one of many seen in the tidal sand flats at the end of Renton Rd, near the airport. It grew here 100,000–200,000 years ago. When it died it fell into a swamp and was buried and preserved in wet peat that has recently been eroded away by the sea.

⊙ The 1.5-metre-diameter broken-off stump of a once majestic kauri tree that grew at Ihumātao 100,000–200,000 years ago when sea level was lower than now. After the tree died the trunk was blown over. The stump and roots remained in place and were buried by black peat (seen here overlying the broken-off stump) in an ancient swamp. The peat in the lower part of the cliff is overlain by lighter-coloured tuff that was erupted from nearby Maungataketake Volcano some time later.

smaller, more branching trees than the earlier near-pure stand of kauri. Leaves were stripped from the branches by the first few ash showers and are today found fossilised in the lower deposits. These tell us that the forest was dominated by rimu, miro, hīnau, tānekaha and a few kauri. Although the initial base surge eruptions do not appear to have felled many of the trees, their standing trunks, which are buried, are in places partly pushed over at crazy angles. Many of their branches broke off under the extra weight of built-up wet ash and now lie fossilised in the lower layers around the in situ tree stumps and trunks. Volcanic hailstones (accretionary lapilli) are common in the layers of tuff and show that base surge pyroclastic flows were a significant part of the early explosive eruptions.

The fossilised kauri stumps and logs that are preserved in the peat and protrude from beneath the sand in the foreshore are nearly as fresh and hard as modern wood. This is because the wood has remained waterlogged in the peat deposit since burial. The stumps and branches of the younger forest preserved in the upper or thinner parts of the layered tuff in the cliffs have obviously not remained waterlogged, as their wood is now mostly badly rotted or completely gone, leaving hollow moulds behind.

These two fossil forests tell us that kauri still thrived in the Auckland area during the colder climate of the ice ages. They provide an unequalled opportunity to glimpse the primeval forests that grew over metropolitan Auckland at a time when many of our volcanoes were erupting and many thousands of years before they were decimated by human colonisation. Pre-European Māori referred to the fossilised kauri at this locality as hora-ko, meaning 'the scattered digging sticks'.

Places of interest around Ihumātao Fossil Forest:
See page 277.

Te Pūkaki Tapu-o-Poutukeka/ Pūkaki Lagoon

Land status: The crater floor is Te Ākitai Waiohua iwi land, and most of the inner slopes of the tuff ring are owned by Auckland Council as a future reserve, but currently they are leased for farming and not open to the public. Most of the crest and outer slopes of the tuff ring are in private properties, except for the roads.

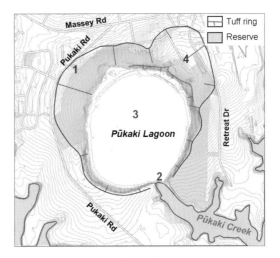

Places of interest around Pūkaki Lagoon:

1. View Pūkaki Lagoon crater from crest of its tuff ring (best between 40 and 42 Pukaki Rd).
2. Location where Pūkaki tuff ring was breached by rising sea level 8000 years ago. It was dammed in the 1920s.
3. Flat floor of crater is underlain by lake and intertidal lagoon sediment.
4. Large slump in inner wall of tuff ring.

Note: Although much of the inside of Pūkaki Lagoon crater is publicly owned for a future reserve, it is currently not open to the public.

⊙ View south over Pūkaki Lagoon explosion crater and surrounding tuff ring with the end of Auckland Airport runway in the top right. *Photo by Alastair Jamieson, 2018*

283

Geology

Pūkaki Lagoon is one of the best preserved of the explosion craters and surrounding tuff rings in the Auckland Volcanic Field. It was formed at least 54,000 years ago when the climate was cooler than today and sea level was lower. Pulsating wet explosive eruptions and base surges resulting from the interaction of rising molten magma with cold groundwater created a 100-metre-deep, 500-metre-diameter crater surrounded by a 20–40-metre-high tuff ring.

The explosion crater slopes are relatively steep with evidence of slumping of tuff back into it soon after eruption. The outer slopes of the tuff ring are more gently sloping and extend up to 500 m from the rim. The ash is weathered to a rich red soil that in places is being used for horticulture. Magma supply stopped before all groundwater was exhausted and no fountaining or lava-flow eruptions occurred. Ferdinand von Hochstetter marked a small scoria cone inside Pūkaki Lagoon crater on his 1864 map, but this was clearly a mistake and he may have been confused with Crater Hill next door.

Drilling in the floor of Pūkaki explosion crater has shown that, following the eruptions, the explosion crater (70 m below present sea level)

filled with fresh water and became a lake. No streams flowed into the lake and it periodically overflowed across the lowest part of the crater rim in the south. Little soil or sand washed into the lake either, and the main component of the sediment that slowly accumulated on its floor was the siliceous shells of diatoms (phytoplankton). Every so often volcanic ash from a distant eruption landed in the lake and sank to form a distinctive layer on its floor. Coring of this sediment has revealed the presence of: at least 14 black basaltic ash layers erupted from nearby Auckland volcanoes; 25 lighter grey, thin andesitic ash layers erupted from Ruapehu, Tongariro and Taranaki volcanoes; and 16 cream to white rhyolitic ash layers that were erupted from the Taupō and Ōkataina caldera volcanoes. Much of the 1.5 m thickness of basaltic ash encountered at 56–58 m beneath the surface was probably erupted from adjacent Crater Hill about 31,000 years ago. Fine scoria mantling the east side of Pūkaki tuff ring probably also came from Crater Hill fire-fountaining.

The lake sediment also contains abundant fossil pollen and spores that record the changing vegetation of the surrounding Manukau lowlands as the climate fluctuated between cool and

◕ Looking north over Pūkaki Lagoon and tuff ring in 1949. Rangitoto is in the distance. *Whites Aviation Collection, Alexander Turnbull Library*

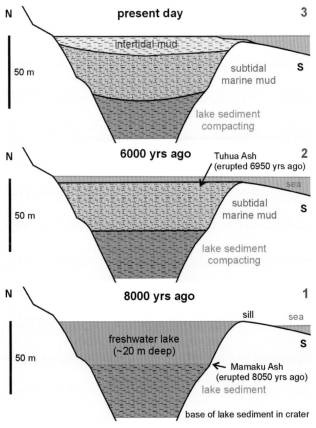

present day 3

N

intertidal mud

50 m

subtidal S
marine mud

lake sediment
compacting

6000 yrs ago Tuhua Ash 2
(erupted 6950 yrs ago)

N

sea

subtidal S
marine mud

50 m

lake sediment
compacting

8000 yrs ago 1

N

sill sea

freshwater lake S
(~20 m deep)

50 m
← Mamaku Ash
(erupted 8050 yrs ago)
lake sediment

base of lake sediment in crater

◉ North–south cross-section through
Pūkaki Lagoon crater showing its changing
geography through the last 8000 years.
The freshwater lake (1) was breached by
rising sea level 8000 years ago. The tides
brought in suspended mud that filled
the tidal lagoon to become tidal flats
by 6000 years ago (2). Since then the
weight of this mud has compacted the
underlying soft layers of lake sediment
creating space for the accumulation of
a further 10 m of intertidal mud (3).

warm, dry and wet, during the ups and downs of
the Last Ice Age cycle. As sea level rose after the
Last Ice Age it reached the height of the overflow
sill and the crater was breached 8000 years ago.
At the time the lake was still about 30 m deep
and initially a deep saltwater lagoon was created.
Every time the tide came in, it carried with it
suspended mud, which sank once inside the quiet
lagoon, quickly filling the crater up with 25 m
of marine mud. By 6000 years ago, Pūkaki had
become entirely tidal mud flats, which gradually
accumulated more sediment that became colonised
by salt marsh and mangroves around the fringes.

Human history
The name for this volcano is a shortened version
of Te Pūkaki Tapu-o-Poutukeka, meaning 'the
sacred fountainhead of Poutukeka', a spring of
ceremonial importance at the head of the lagoon.
In the late 1920s, the tidal inlet was dammed and
the lagoon floor transformed into farmland and
Henning's Māngere Speedway. Ownership of this
reclaimed lagoon floor was returned to local iwi
in the 1990s. Most of the farmed inside slopes of
the crater were purchased for public reserve by
Manukau City Council in 2008 but are not yet
open to the public.

Crater Hill

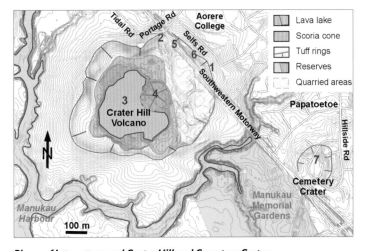

Places of interest around Crater Hill and Cemetery Crater:

1. View Crater Hill from above the Southwestern Motorway.
2. View Crater Hill from gateway on Portage Rd.
3. Freshwater lake in centre of Crater Hill's crater; in public ownership but not accessible.
4. Stump of small quarried-away scoria cone.
5. Site of former small explosion crater.
6. Highest point of Crater Hill tuff ring.
7. Hillside South Park, 151–153 Hillside Rd. Centre of former Cemetery explosion crater.

⊙ View of Crater Hill from the southwest in 2017, clearly outlining the scalloped crest of its tuff ring. The flat circular stump of the quarried scoria cone can be seen on the far, eastern shore of the crater lake. *Photo by Alastair Jamieson*

Land status: The majority of Crater Hill volcano is privately owned farmland. The Southwestern Motorway passes through the eastern side of its tuff ring.

How to see it: A view of the crater can be obtained from the entrance gate opposite 261 Portage Rd, on the west end of the motorway overbridge. The best views of the whole volcano are from out of the windows on the north side of aeroplanes landing to or taking off from the east of Auckland Airport.

Geology

Crater Hill Volcano lies between Papatoetoe and Auckland Airport. In the mid-1980s, the Southwestern Motorway between Onehunga and Wiri was carved through the eastern side of its tuff ring.

This volcano erupted about 30,500 years ago and is named for its spectacular circular crater. Early eruptions were of the wet explosive style, throwing out large volumes of pulverised wet ash interspersed with fine scoria from localised fire-fountaining from a second vent. Together these built up a 25–35-metre-high, 800-metre-diameter tuff ring. The highest point is 40 m ASL on the east side of the motorway. The tuff ring crest is scalloped on the inside. These are believed to be the head scarps of a number of small slumps where wet ash slid back into the crater as the tuff ring was still forming. Early dry eruptions built a wide, low scoria mound inside the northeastern part of the explosion crater but almost all signs of it were buried by later eruptions or have been quarried away. A small, 50-metre-diameter wet explosion crater erupted for a short time on the northern rim of the tuff ring but it has been destroyed by the motorway roadworks.

Once the supply of groundwater was exhausted, lava welled up inside the crater, partially filling it with a lava lake. The surface of the lake cooled and crusted over while staying molten underneath. A little later the lava partly withdrew back down the volcano's throat and the thin, solid basalt crust of the lake was sucked back with it. As the lava withdrew, the basalt crust on the edge around the inside of the tuff ring was left behind like the dirty ring around the inside of a bath when the water is let go. The crust began to reform but there was another phase of withdrawal and a second, lower ring of basalt was left around the inside of the explosion crater. As the lava drained out the collapsing basalt crust became heaped up as a mound of broken basalt blocks

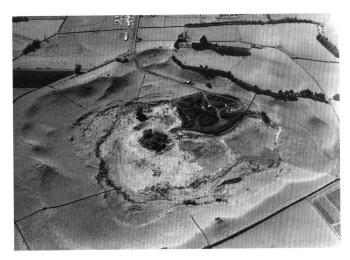

◐ View west over the crater lake in Crater Hill showing the line of basalt rock that marks the former level of a lava lake that welled up inside before later draining back down the volcano's throat.

◑ View east across Crater Hill Volcano prior to its being damaged by the Southwestern Motorway, which was constructed through its eastern part in the 1980s. Here, in 1958, quarrying is removing the last of the small scoria cone that once stood inside the main explosion crater. Note the small explosion crater on the northeastern rim (distant) of the tuff ring, now also gone. *Whites Aviation Collection, Alexander Turnbull Library*

in the centre of the crater. This heap forms an island in the middle of the present-day freshwater lake. In two places on the southern side of the explosion crater the molten lava flowed out from beneath thicker basalt crust on the periphery of the lake, creating several unusual lava caves. During the time of the lava lake, fiery explosive eruptions broke out on its eastern edge, building a sizeable, 25-metre-high scoria cone on the cooled solid basalt crust on the lake surface.

Human history

No Māori name survives for this volcano, although in the last few years the collective name for all the crater volcanoes in this vicinity – Ngā Kapua Kohuora – has been applied to Crater Hill. The inner and outer slopes of the tuff ring were extensively cultivated in pre-European times and house sites are known on the tuff ring crest and inside the shelter of the large crater and near the freshwater lake. This would have been a strategically important site because it was close to one of the major canoe portages between the Manukau and Waitematā harbours.

The two best lava caves inside the crater are known as Self's Lava Cave, named after the family who has owned this property since the 1920s, and Underground Press Lava Cave, which is reputed to have been used for clandestine subversive publishing during the Second World War. Today, the ephemeral freshwater lake in the bottom of the crater is present in winter and almost dry in summer. This lake is a water reserve and for many years was used as a supply of fresh water for Dominion Breweries in Papatoetoe from the 1930s.

The near-central scoria cone was quarried down to a flat stump in the 1940s–50s. The scoria-rich tuff ring in the east was also extensively quarried during the latter part of the 20th century because of its high scoria content. In the last decade or so, the large tuff ring quarry has been filled with cleanfill, approximately restoring the original shape of the tuff ring, but only the flat, quarried-off stump of the scoria cone remains, next to the lake.

Kohuora Crater

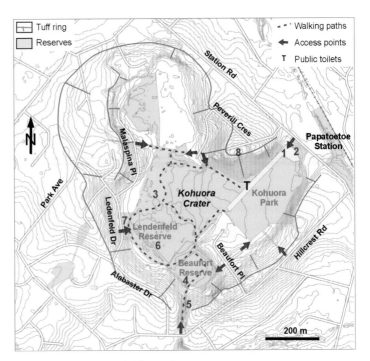

Places of interest around Kohuora Crater:

1. Main entrance and car park for Kohuora Park.
2. Papatoetoe Rugby League clubhouse.
3. Walkway and cycle paths; boardwalks through swamp.
4. Beaufort Reserve small explosion crater.
5. Overflow stream cut through tuff ring.
6. Lendenfeld Reserve and wetland. Crater is filled with lake and swamp sediment.
7. Small playground, 26 Lendenfeld Drive.
8. Highest point on tuff ring, 35 m ASL.

⊙ View south over the L-shaped Kohuora explosion crater and surrounding irregular tuff ring. *Photo by Alastair Jamieson, 2009*

Land status: The floor of the east and much of the southwest and northwest lobes are in reserve administered by Auckland Council, as are some of the inner slopes of the crater in the east. The remainder of the crater and tuff ring is in private houses or roads.

What to do: Walk or cycle around the many paths on the crater floor.

Geology

Kohuora is an L-shaped crater produced by wet explosive eruptions from four and possibly more vents as rising magma interacted with voluminous groundwater. As a result, this is one of the few explosion crater volcanoes in Auckland that do not have a simple circular shape. The explosive eruptions threw out vast quantities of pulverised light-coloured pumiceous silt and Waitematā Sandstone, which underlies most of the Manukau lowlands, and lesser amounts of darker volcanic ash. This built up an encircling rim of bedded tuff, 15–25 m high, around the vents. Kohuora Crater erupted about 34,000 years ago and its tuff ring and crater were mantled by dark scoriaceous ash blown from nearby Crater Hill when it erupted

about 3000 years later. Probably the youngest of the explosion crater vents was in the southwest corner and occupies the low circular depression of Beaufort Reserve. Some of the ash thrown out from there forms the low land now underlying the houses in Beaufort Place.

Originally, Kohuora's complex crater was a lot deeper than today, but following its formation it became a lake that accumulated much sediment. By the time of human arrival the floor of Kohuora Crater was a wetland swamp. A buried pipe beneath the natural overflow saddle drains the basin at the southwest corner from Beaufort Reserve.

Human history

Kohuora is the Māori name applied to this unusual volcano west of Papatoetoe railway

⊙ The small Beaufort Reserve explosion crater, seen here in 2018, was probably formed by the last wet explosive eruptions from Kohuora.

◐ View east over Kohuora crater and tuff ring in 1949 before the floor had been drained and developed into reserve and before houses were built on the tuff ring. *Whites Aviation Collection, Alexander Turnbull Library*

station and refers to the mist that sometimes fills the crater, providing life-giving moisture. Both Charles Heaphy and Ferdinand von Hochstetter recognised that this was a crater of volcanic origin, although Hochstetter inferred that a number of other less prominent depressions nearby were also craters. He labelled them 'the five craters of Kohuora', although today the name applies only to this one – the two he (and not Heaphy) recognised to the east have not stood up to the scrutiny of later geologists and are not believed to have been volcanic craters.

Kohuora has not been modified by any quarrying, but much of the crater floor wetland has been filled with refuse and clay and drained. The eastern lobe of the crater is Kohuora Reserve and Papatoetoe Rugby League Club playing fields. Some of the southern and northwestern lobes are now covered by houses (Beaufort and Malaspina places) and much of the remaining unfilled wetland is set aside as Beaufort and Lendenfeld reserves.

◐ Walkways have been built through and around the Kohuora crater wetlands.

Cemetery Crater

⊙ Cemetery Crater is the circular depression with surrounding raised ridge beyond the Manukau Memorial Gardens in Papatoetoe, seen here in 1949. Today the Southwestern Motorway runs right across the middle of the photograph. *Whites Aviation Collection, Alexander Turnbull Library*
⊙ View from the southeast over the site of Cemetery Crater (centred on clump of trees in middle) and surrounding tuff ring. It was named after the Manukau Memorial Gardens cemetery now on the west side of the motorway. The crater lake in Crater Hill is in the distance. *Photo by Alastair Jamieson, 2018*

Land status: All of Cemetery Crater is within a residential subdivision. Publicly accessible are the roads and Hillside South Park (151–155 Hillside Rd), which is close to the centre of the former crater.

Cemetery Crater is located 1 km southeast of Crater Hill and 1 km south of Kohuora Crater. It was a shallow, 200-metre-diameter explosion crater surrounded by a 10-metre-high tuff ring, which in the southwest has been breached by a stream. It was first recognised in the 1960s by geologist Ernie Searle who named it after the nearby Manukau Memorial Gardens cemetery in Puhinui Rd. Since then the two have been separated by construction of the Southwestern Motorway. At the time of residential subdivision the tuff ring and crater were recontoured to make more gently sloping sections. There are no rock exposures to be seen and we do not know when this volcano erupted.

Places of interest around Cemetery Crater:
See page 287.

Ash Hill Crater

◉ View south over the site of Ash Hill explosion crater, which lies hidden beneath industrial buildings and yards (centre) on the east side of Ash Rd (on right), Wiri. *Photo by Alastair Jamieson, 2018*

Land status: The site of Ash Hill is privately owned industrial subdivision. From the roadside you can view the site in the block of land between Ash, Oak and Wiri Station roads.

Ash Hill was a small, 125-metre-diameter explosion crater with a low surrounding tuff ring, all of which has now been removed during industrial development. The explosive eruptions appear to have been phreatic (steam) blasts, at times including some fragmentation and eruption of chilled basalt lava. Drilling investigations at the time of subdivision showed that the crater was about 30 m deep and is filled with mud and peat which accumulated on the floor of a freshwater lake that later became a swamp. Ash Hill is at the northeastern end of a line of three volcanoes in this area – McLaughlins Mountain, Wiri Mountain and Ash Hill. The magma that erupted at each centre probably rose to the surface along the same northeast–southwest-oriented hidden fault line. Ash Hill erupted just before Wiri Mountain, about 31,000 years ago. The crater was first recognised by geologist Ernie Searle in the 1960s. He named it after nearby Ash Rd with its nearby ash trees (not volcanic ash). It has no known Māori name.

◔ Prior to industrial subdivision, Ash Hill had a swampy crater surrounded by a low circular tuff ring, as seen here in a vertical aerial photo taken in 1959. *Courtesy of LINZ historic aerial photo archive*

◑ An exposure of tuff in 2005 that once formed part of Ash Hill's tuff ring. Note the angular blocks and fragments of a wide range of sediment types that were thrown out by the explosive eruptions along with basaltic ash and lapilli. Photo 1 m wide.

Places of interest around Ash Hill Crater:
See page 299.

Te Manurewa-o-Tamapahore/Matukutūruru/ Wiri Mountain

◉ View north over the quarried-out site of Wiri Mountain's scoria cone. Only a remnant of the lower northern slopes containing Wiri Lava Cave remains (top left). *Photo by Alastair Jamieson, 2018*

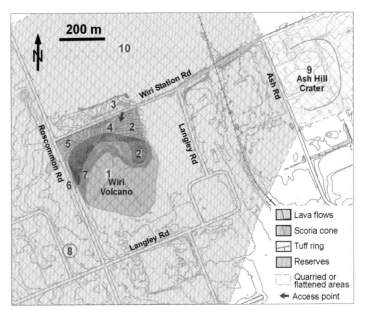

Places of interest around Wiri Mountain and Ash Hill Crater:

1. Flat earth-filled quarry floor that is the site of most of the quarried-away scoria cone.
2. Wiri Lava Cave Scientific Reserve, surrounded to the south by scoria cliffs.
3. Toe of Wiri Lava Cave Scientific Reserve on north side of Wiri Station Rd.
4. Rumney cottage (Jacaranda House, built about 1928) made of basalt rock and concrete.
5. Pre-European stone garden area on former lower slopes of the cone and lava flows.
6. Roscommon Rd cutting showing bedded tuff overlain by basaltic lava flow.
7. Old quarry face with lava flows butting up to a slump scarp in the tuff ring.
8. Site of former steam eruption crater, now destroyed (on private land).
9. Site of Ash Hill explosion crater, now destroyed (on private land).
10. Former Wiri North Quarry where large Wiri lava-flow field was removed and backfilled with rock from excavation of Waterview Tunnel.

Land status: The remains of the lower northwest slopes of Wiri Mountain are administered by the Tūpuna Maunga Authority. Wiri Cave Scientific Reserve underlies the lower northeastern slopes and is managed by the Department of Conservation. The remainder of the volcano's footprint is in private property.

Geology

Wiri Mountain erupted approximately 30,500 years ago. Remnants of a tuff ring seen in the quarry walls show that it began with wet explosive eruptions that would have produced a large crater and surrounding tuff ring. The dry fountaining eruptions that followed built a large 60-metre-high scoria cone that filled the explosion crater and subsequent outwelling of lava flows buried the low tuff ring.

Lava flowed outwards in all directions from the base of the cone creating an apron that completely encircled it and extended 1 km to the north and 1.5 km to the south. Most of these flows have been quarried away in recent decades. In the north, the lava flows were 10–18 m thick. To the southwest, some of Wiri's lava flows spread across earlier flows from McLaughlins Mountain and reached the base of that volcano.

Human history

Several Māori names have been applied to this volcanic cone, with Wiri Mountain the most used today, derived from Wirihana Takanini, the Māori chief who sold the hill last century. Another name accepted by local iwi is Te Manurewa-o-Tamapahore, source of the name Manurewa used for the suburb just to the south. The mountain was terraced and used as a fortified pā by pre-European Māori.

These days there is not much left of the original volcano – just the lower northern slopes of the scoria cone, part of which contains the best lava cave in New Zealand: Wiri Lava Cave

● View southeast over the remains of Wiri Mountain showing the extensive pre-European terracing. The railways quarry is in the left side of the cone. The road on the right is Roscommon Rd. *Jack Golson, mid-1950s, University of Auckland Library*

● Wall of old Wiri Mountain quarry in 2018 showing portion of the tuff ring (right) unconformably overlain (left) by a series of dark lava flows. The sloping unconformity between them is probably a slump scarp where some of the unstable wet tuff slid back into the explosion crater during the eruption.

● View northwest over the remains of Wiri Mountain in 1958, showing the branch railway line (bottom right) from the main trunk line leading into the New Zealand Railways quarry. *Whites Aviation Collection, Alexander Turnbull Library*

Scientific Reserve (see following section). Quarrying of this cone began in a small way in 1859 to aid in the construction of the Great South Road. The main trunk railway line runs alongside, and after the acquisition of the cone by New Zealand Railways in 1915, quarrying activities moved up a notch with the scoria from Wiri Mountain used for railway ballast all the way south to Ōhākune. The deep quarrying into the base of the scoria cone exposed some of the volcano's plumbing that fed lava up fissures and pipes to erupt as flows from around the cone's lower flanks. The large hole that was left after the scoria cone had been quarried away has been filled and is being used by industry. So are the many flat areas that were left after the lava flows were quarried for aggregate.

Wiri Lava Cave

Wiri Lava Cave is New Zealand's longest (200 m) and best example of a lava cave. It is protected within a scientific reserve beneath the only significant remnant of Wiri's quarried scoria cone. A permit from the Department of Conservation is required for entry. Wiri Lava Cave is also rare in New Zealand as the one of only two known accessible caves that formed within a scoria cone.

The upper half of the cave was formed by lava that rose up inside Wiri Volcano's throat and pushed its way through the northern flanks of the scoria cone to emerge as a lava flow. As the lava passed through the scoria, some of it cooled to form a thin lining of glassy basalt on the inside of the conduit and this is now the

5–20-centimetre-thick walls of the cave. This upper part of the cave has a 4–6-metre-high ceiling with a unique Gothic arch cross-section. This shape was produced when lava began draining out of the tube and the weight of the surrounding scoria caused the still soft lining rock to sag inwards. Part-way down Wiri Lava Cave, the tube ceiling becomes lower and the walls are much thicker and more solid basalt. This is where the lava emerged from the base of the scoria cone and from here on the cave was the feeder tube inside a lava flow. The lower part of the cave within the flow passes under Wiri Station Rd (crest of rise) and into the scientific reserve land on the other side.

The survival of Wiri Lava Cave is a miracle and a tribute to the New Zealand Historic Places Trust, the New Zealand Speleological Society and, particularly, the Geological Society of New Zealand, who fought for nearly 30 years to save it from destruction during the quarrying of Wiri's scoria cone. The battle heated up in 1970 when the New Zealand Historic Places Trust won its case before the Planning Tribunal to have the cave scheduled for protection on the Manukau District Scheme. The quarry was run by New Zealand Railways and countless Ministers of Railways were lobbied to make sure the cave was not quarried away. It was not until 1998, after a major report from the Parliamentary Commissioner for the Environment calling for action, that the land was transferred to the Department of Conservation and declared a scientific reserve.

◐ Inside Wiri Lava Cave with its classic Gothic arch cross-section. The ropey pahoehoe floor is the remnants of the last lava flow that did not fully drain out of the tube. *Photo by Alastair Jamieson, 2010*
◑ Plan and cross-section of Wiri Lava Cave.
◑ Wiri Lava Cave lies within the strip of unquarried, dark vegetated volcano running near the middle of this 2018 photo. It passes under Wiri Station Rd and beneath the grassed reserve land on the north (foreground) side. *Photo by Alastair Jamieson*

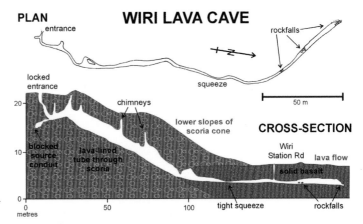

Matukutūreia/
McLaughlins Mountain

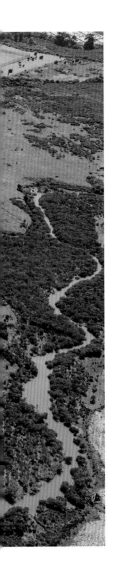

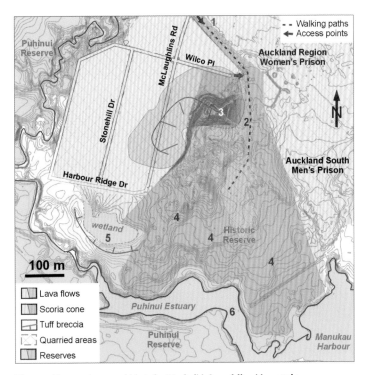

⊙Matukutūreia/McLaughlins Mountain reserve viewed from the west in 2018 showing pyramidal scoria cone remains (top left), the humpy lava-flow field and the arcuate phreatic explosion crater in the foreground. *Photo by Alastair Jamieson*

Places of interest around Matukutūreia/McLaughlins Mountain:

1. Main entrance gate.
2. Site of historic pumphouse.
3. Crest of pyramidal remnant of scoria cone and site of former water reservoir.
4. Stone heaps and rows – remains of pre-European gardens on lava-flow field.
5. Irregular steam explosion crater with wetland surrounded in the southwest by an arcuate ridge of tuff breccia. Lava flows entered the crater from the north side.
6. Southwestern interceptor sewage pipeline bridge.

Land status: The remnant cone and lava flows with stonefield gardens are a historic reserve managed by the Department of Conservation. The rest of the volcano's footprint is in private industrial development and roads.

What to do: Climb to the top of the scoria cone remnant for the view and to see the remaining archaeology. Walk around the Matukuturua pre-European stonefield gardens.

Geology

This is the southernmost volcano in the Auckland Volcanic Field and contains evidence of three styles of eruption. If there was an initial explosive phase, the crater and tuff ring are now buried. A large scoria cone was built by fountaining eruptions and voluminous lava flowed out from around its lower slopes and spread out in all directions to create an apron of lava flows. Lava pouring out from the southern side carried away some of the scoria cone, creating a U-shaped breached crater. Mounds of rafted scoria give the flows on this side a rough, irregular surface. The last phases of eruption from the cone's crater were fiery explosive eruptions of more pasty, less gaseous lava that produced numerous irregular twisted

and elongate bombs and ragged lumps of partly welded scoriaceous basalt that caps the cone.

Near the end of the eruption sequence, it appears that a dike of magma intruded up into the wet rocks out to the southwest of the cone. This heated the groundwater and resulted in one or more large phreatic (steam) explosions through the edge of the early lava flows, creating a crater surrounded by a ring of tuff breccia. The remaining arcuate tuff breccia ridge around the southwest side of the crater is a structureless heap of angular blocks of sedimentary rock and basalt in a matrix of sandy mud, all of which was broken up by the explosive blasts. Some of the last lava erupted from the base of the scoria cone appears to have flowed into the northern side of the crater creating its irregular margin.

⬆ View west from the lower slopes of McLaughlins Mountain showing the curved tuff ring remnant and enclosed freshwater wetland in the remaining part of the phreatic explosion crater.

This explosion crater became a freshwater pond that gradually filled with sediment and in recent times has become a seasonally flooded swamp with remnant native wetland plants.

Human history

Like all of Auckland's scoria cones, Matukutūreia was terraced and used as a fortified pā by local Māori in pre-European times. The rich, volcanic-derived soil on the surrounding lava flows was intensively used by them for cultivation and growing kūmara and a few other crops. The naturally stony surface of the flows was modified as part of the gardening activities. Larger rocks were heaped into rows or mounds, often on top of natural rock outcrops. In places there are flattened, rectangular enclosures that were probably the sites of whare (dwellings). The 43-hectare historic reserve was created to protect the remnants of the Matukuturua Stonefield gardening site.

The European name, McLaughlins Mountain, came from the long-time European owners. From 1929 to the 1960s, the Borough of Papatoetoe obtained water from a bore into scoria at the foot of McLaughlins Mountain. The water was stored in a reservoir on the summit of the cone. The first quarrying of the scoria cone began in a small way in the 1850s to supply metal for the nearby Great South Rd. Quarrying on a larger scale did not commence until 1960, first removing all of the cone, except a benched pyramid that was retained to support the reservoir. As the scoria resource ran out, quarrying moved into the surrounding lava-flow field and removed most of it except that which is now in reserve. The reservoir and pumphouse were removed in 2011.

○ This brick pumphouse on the southern side of McLaughlins Mountain in 1962 was used to pump fresh water from an adjacent water bore up to the reservoir (now removed) on the top of the scoria cone as part of the water supply for Papatoetoe Borough. *Whites Aviation Collection, Alexander Turnbull Library*
○ View of Matukutūreia/McLaughlins Mountain from the southwest in 1952 before its profile was forever destroyed by quarrying. Note the arcuate phreatic explosion crater lake in the front left. *Whites Aviation Collection, Alexander Turnbull Library*

Puhinui Craters

⊕ View northeast over Puhinui Craters and their small surrounding tuff rings. Pond Crater is in the top left, Arena Crater is in the bottom right and Eroded Crater is centred on the eroded, scrub-filled valley beyond Arena Crater. *Photo by Alastair Jamieson, 2018*

❯ Near the entrance to
Puhinui Reserve is Pond Crater
surrounded by a low tuff ring.

Land status: All three craters occur within Puhinui Reserve (end of Price Rd), managed by Auckland Council.

What to do: Enjoy a walk, mountain bike ride or picnic in a rural setting.

Puhinui Craters were recognised as volcanic features within the Auckland Field only in 2011. They consist of three small craters (150–250 m in diameter) surrounded by low tuff ring cones. Pond Crater, near the entrance to the reserve, is breached to the west and has a small farm dam controlling water level in the crater. Arena Crater is breached to the east and its swampy crater floor has been drained and slightly modified to form an equestrian arena. It has a flat floor because the crater has since filled up with peat and lacustrine sediment. Eroded Crater is the hardest to recognise as it has been eroded out by a small stream flowing from west to east through its middle, but the raised rim of tuff is still present on either side. The age of eruption of Puhinui Craters is unknown although they possibly erupted at the same time as their neighbour, Matukutūreia, across Puhinui Stream.

All three of these small craters have elliptical shapes with parallel northeast-trending axes, which suggest the magma came up as narrow intrusional dikes along two parallel northeast-aligned faults and the eruptions were fissure-like. Most of the blasts were probably phreatic (steam), produced by groundwater that had been superheated by the rising magma. Some blasts may also have included some cooled and fragmented basalt from the surface of the magma.

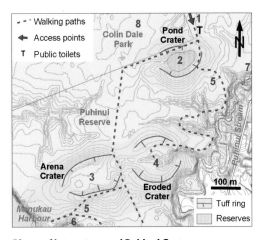

Places of interest around Puhinui Craters:

1. Car park at entrance to Puhinui Reserve.
2. Pond Crater.
3. Arena Crater, used for showjumping and dressage.
4. Eroded Crater.
5. Walking tracks and mountain bike trails.
6. Coastal wildlife area for birds, such as fernbird, black stilt, NZ dotterel and wrybill.
7. Puhinui Stream walkway.
8. Colin Dale Park – new home for Auckland's speedway races.

Historic basalt buildings of Auckland

Most of the early British colonists arriving to settle in Auckland came from a heritage of stone buildings with slate roofs. So it was not surprising that there were early attempts to use local stone to recreate the building styles they were familiar with back home.

Auckland's local sandstone was too soft for building and the hard greywacke rocks were too intensely fractured. This left the basalt from Auckland's lava flows as the only handy rock that might be utilised, but in many places it too was closely fractured and a hard material to work for intricate and precise facing stones. Mortar to cement the blocks together was initially derived

from lime-burning of cockleshell deposits around the shores of the Waitematā Harbour.

One of the earliest sources of basalt to be used was from Rangitoto Island, where small cobbles were collected from the foreshore, ferried across to the growing town and used for rubble-walled buildings. Larger blocks were prised from the lava flows and worked by skilled stonemasons, such as Benjamin Strange. Rangitoto basalt can be seen in a number of 1850s and 1860s Auckland buildings. The most prominent source of basalt for building, especially after the 1880s when the prison quarry opened, has been the thick lava flows from Mt Eden.

1. The Bluestone Room, 9–11 Durham Lane, is the oldest stone building (c. 1861) left in the heart of Auckland City and one of the few remaining mid-19th-century warehouses. It has had many uses over the years, most recently as a local tavern.

2. St Paul's Anglican Church, 28 Symonds St, was built in 1895 using grey basalt from Auckland and Melbourne set off with cream Oamaru Stone facings. The chancel was not added until 1936–37.

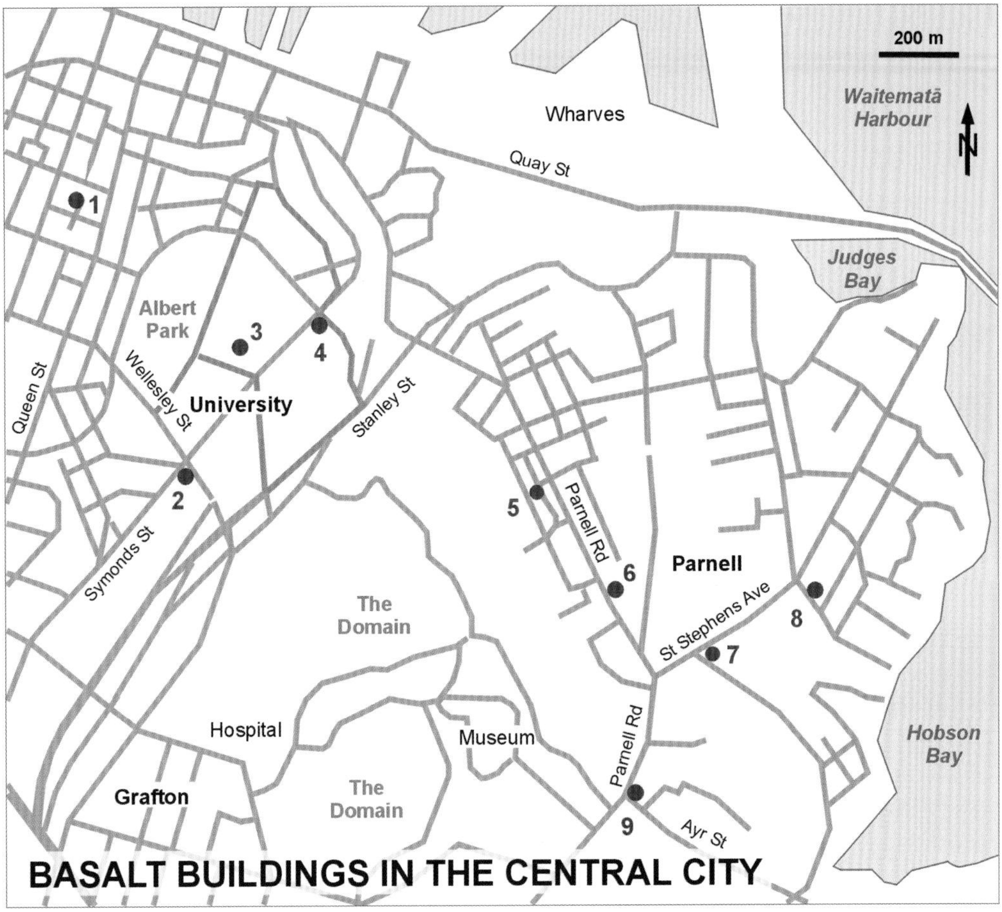
200 m
Waitematā Harbour
N
Wharves
Quay St
Judges Bay
Albert Park
University
Wellesley St
Queen St
Stanley St
Symonds St
Parnell Rd
Parnell
St Stephens Ave
The Domain
Hospital
Museum
Grafton
The Domain
Parnell Rd
Ayr St
Hobson Bay
1
2
3
4
5
6
7
8
9
BASALT BUILDINGS IN THE CENTRAL CITY

3. Albert Barracks Wall, Alfred St, University of Auckland, was constructed by Māori stonemasons between 1846 and 1852, finally enclosing about 10 hectares. Just this small section of wall remains.

5. Stonemason's House, 27 Falcon St, Parnell, was built in 1863 by Auckland's best-known stonemason, Benjamin Strange, for his own use. The rubble has been plastered over and painted, while the stone quoins remain exposed.

4. St Andrew's Presbyterian Church,
2 Symonds St, is the oldest stone church in New Zealand and the oldest surviving church in Auckland. The earliest part was constructed between 1847 and 1850 using basalt quarried from a Mt Hobson lava flow near Newmarket. The steeple was a later (1880s) addition.

6. Whitby Lodge (Beaufort House),
330 Parnell Rd, Parnell, is named after an early owner and is one of the few surviving stone colonial dwellings in Auckland. The wooden part at the back was built first and the stone front portion was made from Mt Eden basalt in 1874.

7. The Deanery, 17 St Stephens Ave, Parnell, was designed by Frederick Thatcher and built by Benjamin Strange in 1857 with Rangitoto basalt rubble walls and dressed stone quoins. It was initially lived in by Bishop Selwyn.

8. Benjamin Strange's house, 4 Takutai St, Parnell, was designed by Frederick Thatcher and built by Benjamin Strange in 1859 using basalt rubble walls and dressed stone quoins from Rangitoto. Commissioned by Bishop Selwyn for the Anglican Church and rented by Strange until he moved to his own house in Falcon St in 1863, it was subsequently occupied by John Kissling, first archdeacon of St Mary's.

9. Kinder House, 2 Ayr St, Parnell, was designed by Frederick Thatcher and built in 1856–57 for Dr John Kinder, first headmaster of the Church of England Grammar School (across the road). It was built by Benjamin Strange using basalt from an unknown source. The walls are made of random rubble held together with mortar and the facings are of large squared blocks of basalt.

10. The Merksworth Castle, 253 Hurstmere Rd, Takapuna, was built in 1926 for John and Penelope Algie. Its 1-metre-thick walls are made of basalt quarried from the south shore of Lake Pupuke.

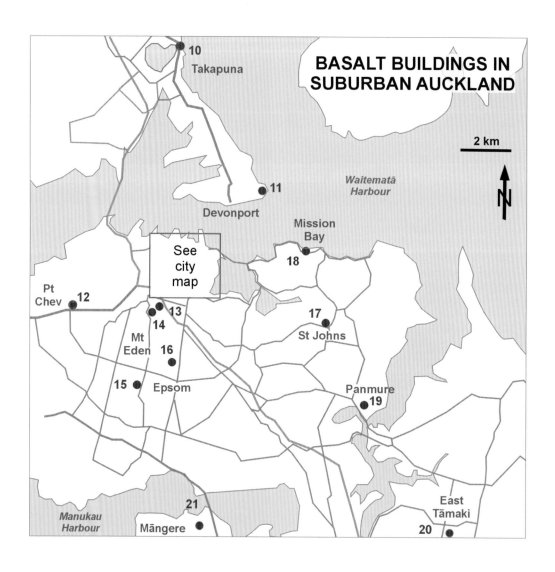

314

11. Barracks' Kitchen, North Head, DOC Visitor Centre. This rectangular building was built in 1885 from basalt taken from the nearby Devonport foreshore (Mt Victoria lava flows). It was built for the New Zealand Armed Constabulary and used by the army and navy until 1996 when it was passed over to the Department of Conservation.

12. Potter Cottage, MOTAT, Western Springs, was built in 1854 as a rural dwelling on William Potter's farm in Manukau Rd (now Alexandra Park Raceway). It was shifted stone by stone to the Museum of Transport and Technology site in 1969.

13. Mt Eden Prison, Lauder Rd, was built by prisoners starting in 1883 and completed in the early 20th century. The basalt was quarried on site from a thick Mt Eden lava flow. Its exterior was modelled on castles with classical or Gothic architecture.

14. CAC building, 26 Normanby Rd, Mt Eden. Built in 1922, as part of the Colonial Ammunition Company (CAC) industrial complex; the associated 1916 shot tower, used in the manufacture of lead shotgun cartridges, still stands behind. Now used as a bar and eatery.

15. Blinkbonnie, 753 Mt Eden Rd, Mt Eden. A mid-Victorian two-storey cottage of simple Georgian design built *c*. 1862 for English settler Robert Joughin, who farmed 100 acres.

16. Epsom Community Centre stone houses, 200–202 Gillies Ave. Both houses were built from Mt Eden basalt quarried across the road by the Epsom Roads Board in the 1900s and 1930s.

17. St John's College kitchen, 202 St Johns Rd, Meadowbank, was built between 1845 and 1846 for the Anglican Theological Training College of St John the Evangelist. It is made of Rangitoto foreshore basalt and named the 'Waitoa Room' after the first Māori to be ordained – Rota Waitoa. Built with beach sand mortar and Waitematā Sandstone facing stones.

18. Melanesian Mission House, Mission Bay, was built in 1859 and is the last remaining structure from the St Andrew's College Anglican missionary training school (1858–67), sited beside the sea so the Melanesian students would feel at home. Designed by Reader Wood as the college's dining room and kitchen, the mission house was constructed by Benjamin Strange out of basalt collected from the Rangitoto Island foreshore. Today, it is used as a restaurant.

19. Panmure stone cottage, 1 Kings Rd, Panmure, was built for Mr Peter Brady about 1855 and, after many owners, was shifted to its present site in 1972. It has been restored and refurnished with items from the 19th century and is periodically open to the public.

20. St John's Church, Hampton Park, East Tāmaki, was constructed from locally quarried basalt in 1862 under the guidance of the Rev. Gideon Smales. While the walls were made of random basalt blocks, rendering and pointing were added to give the impression of ashlar construction in sandstone. The church has been restored since a disastrous fire in 1965. Nearby in Hampton Park are the ruins of Gideon Smales' stone stables, built and burned down in the 1870s.

21. St James' Church, 39 Church St, Māngere Bridge, was constructed from local basalt from Māngere Mountain. This Anglican church was designed by the Rev. Guyon Purchas, who also supervised its construction between 1857 and 1859. It is the oldest Selwyn-style stone church in New Zealand.

⊙ Western Springs bubble out from the fractures in the longest lava flow from Three Kings Volcano. In 1875, a 200-metre-long dam was built to create the 2-metre-deep Western Springs pond so that a continuous supply of water was available to be pumped up to reservoirs to quench the thirst of fledgling Auckland City. Today the pond and surrounding land is a popular public park. *Photo by Alastair Jamieson, 2009*

Glossary

a'a lava	(pronounced ah-ah) lava flow that moves slowly and has a rubbly or blocky surface
accretionary lapilli	small (1–10 mm diameter) balls of fine volcanic ash that grew in a moist eruption cloud by ash particles sticking together, and then rained to the ground where they can be seen in volcanic ash beds (tuff); also known as volcanic hailstones or chalazoidites
aeromagnetic survey	measuring the magnetic properties of the near-surface rocks with a magnetometer towed behind an aeroplane
alkali basalt	a medium-to-dark grey volcanic rock with 45–49 per cent silica content; the main rock type erupted by Auckland volcanoes
ash	mud and sand-size fragments (less than 2 mm across) of volcanic and other rock that have been fragmented and erupted into the air from a volcano
ASL	above sea level
basalt	dark grey-coloured, fine-grained volcanic rock with 45–52 per cent silica composition
base surge	a superheated cloud of turbulent steam, volcanic gas, ash and rock fragments that is blasted sideways from a vent during wet explosive eruptions and races across the ground at considerable speed; the most dangerous style of eruption from an Auckland volcano
breached crater	a U-shaped crater on a scoria cone with one of its sides rafted away by a lava flow
breccia	rock composed of angular boulders, cobbles or pebbles (fragments greater than 2 mm in diameter)
conduit	pipe or passageway for conveying fluid, such as magma
country rock	the background rock through which the volcanoes erupted
explosion crater	wide, relatively shallow, circular crater remaining after a series of wet explosive eruptions (also known as a maar crater), e.g. Panmure Basin, Pūkaki Lagoon
explosive eruption	a violent wet style of eruption that occurs when molten magma encounters cold water and blasts out a cloud of steam, gas and fragmented rock (also known as phreatomagmatic eruption = steam and magma eruption)

fiery explosive eruption	a violent dry style of eruption of pasty magma powered by gas release and large bursting bubbles throwing out incandescent ragged lumps of magma or aerodynamically moulded bombs (also known as Strombolian eruption)
fire-fountaining eruption (or lava-fountaining)	steady fountaining eruption of fluid magma powered by gas release and building up a scoria cone around the vent (also known as Hawaiian eruption)
fossil	any remain or trace of a plant or animal that has been preserved in rock
geophysical survey	measuring the physical properties of the Earth using a variety of remote sensing techniques
hornito	a small steep-sided cone or chimney of lava spatter built up around a hole in the crusted-over roof of a lava tube within a lava flow
join	in volcanic rocks this is a surface along which the rock breaks due to shrinkage that accompanied cooling of lava and its solidification to rock
lapilli	pebble-sized pieces (2–64 mm across) of volcanic and other rock that have been fragmented and erupted from a volcano; e.g. pebbles of vesicular basalt are called scoria lapilli
lava	molten rock (magma) that has been erupted onto the Earth's surface
magma	molten rock (up to 1200 °C) occurring under the ground; may erupt at the surface (after which it is called lava but is essentially the same composition as the parent magma)
midden	pre-European rubbish dump, usually containing shells, bones, charcoal and burnt oven stones
monogenetic volcano	a kind of volcano that forms during a short-lived eruption of a few months to maybe decades, and never erupts again
olivine	a crystalline mineral composed of magnesium iron silicate that sometimes crystallises in the magma underground and may be present in Auckland's basalt rocks or scoria deposits
pā	fortified Māori village or fort often sited on Auckland's scoria cones; may be defended by ditches, banks and high wooden fences (palisades)
pahoehoe lava	(pronounced pa-ho-ee-ho-ee) lava flow that has a rapidly chilled, rolled-up or ropey but otherwise smooth, glassy surface
paleomagnetic	the record of the direction of ancient magnetisation preserved in iron-rich minerals in rocks
phreatic eruptions	explosive eruptions of steam and fragmented rock
phreatomagmatic eruptions	wet explosive eruptions resulting from the interaction of molten magma with cold water
pit	rectangular-shaped holes that had roofs over them and were used in pre-European times for food storage, particularly kūmara
projectile block	a lump (larger than 64 mm across) of solid rock thrown out from a volcanic vent, mostly during wet explosive eruptions
pumice	solidified frothy volcanic rock formed in violent wet explosive eruptions; usually pale-coloured and rich in silica. Most pumice around Auckland was erupted from the rhyolitic caldera volcanoes of the Taupō–Bay of Plenty region

scoria	lightweight volcanic rock full of holes (vesicular) that was erupted by fire-fountaining or fiery explosions of frothy lava which cooled and solidified as it travelled through the air (sometimes referred to as cinders)
scoria cone	relatively small, usually steep-sided (c. 30 degrees) volcanic cone made of scoria erupted by fire-fountaining and fiery explosive eruptions from one or more vents (also known as a cinder cone), e.g. Mt Eden, One Tree Hill
seismological	referring to study of the vibration or shaking in the Earth caused by earthquakes or man-made explosions
seismometer	instrument used for detecting vibration waves passing through the ground
shield volcano	moderately large, gently sloping (c. 10 degrees) volcanic cone made of overlapping basaltic lava flows erupted from one or more near-central vents, e.g. Rangitoto
slump scarp	a steep bank left behind when a mass of rock and earth slides off downhill
spatter (lava spatter)	fluid fragments of molten lava ejected from a vent during fire-fountaining or fiery explosive eruptions that may flatten, splash and solidify when they land and sometimes weld together to form a small, steep-sided spatter cone; frequently erupted late in the life of an Auckland scoria cone
tephra	solid material that has been erupted into the air by a volcanic eruption and deposited on the ground; includes all fragment sizes – ash, lapilli and blocks
tephrochronologist	a scientist who studies the sequence of volcanic ashes
terraces	levelled areas on the slopes of Auckland's volcanoes; mostly constructed by pre-European Māori and used for dwellings and storage buildings
tuff	volcanic ash that has hardened to become rock
tuff ring	a near-circular rampart of bedded volcanic ash (tuff) built up around an explosion crater (also known as a tuff cone when larger), e.g. Ōrākei Basin
vent	the opening through which volcanic material erupts
vesicular	full of holes (vesicles)
viscous	thick, sticky, resistant to flow, like golden syrup
volcanic bomb	a glob of magma ejected into the air while still molten and often acquiring an aerodynamic shape as it cools during flight
Waitematā Sandstone	bedded sandstone and mudstone strata that underlie most of Auckland's volcanoes

Select bibliography

General references

The general references are listed by year of publication to give readers an idea of the progression of studies that has advanced over time. Although some earlier titles have been superseded by more recent books, all have contributed to our present understanding.

Hochstetter, F. von (1864). *Geology of New Zealand* (C. Fleming, trans. 1959). Wellington: Government Printer, 320 pp.

Special Committee of Auckland Town Planning Association (1928). *Auckland's unique heritage. 63 wonderful volcanic cones and craters. An appeal to save them*, 26 pp.

Golson, J. (ed.) (1957). *Auckland volcanic cones*. Auckland: Historic Auckland Society, 32 pp.

Searle, E. J. (1964, 1981). *City of volcanoes: A geology of Auckland*. Auckland: Longman Paul, 195 pp.

Fowlds, G. M. (1964). *Place names of the volcanic cones and craters of the Auckland isthmus and origin of the English titles*. Auckland: Auckland Town Planning Association, 26pp.

Searle, E. J., & Davidson, J. M. (1973). *A pictorial guide to the volcanic cones of Auckland showing geological and archaeological features*. Auckland: Auckland War Memorial Museum Handbook Series, 27 pp.

Graham, G., & Simmons, D. (eds) (1980). *Māori place names of Auckland: Their meaning and history*. Auckland: Auckland Institute and Museum, 39 pp.

Ballance, P. F., & Smith, I. E. M. (1982). *Walks through Auckland's geological past: A guide to the geological formations of Rangitoto, Motutapu and Motuihe Islands*. Wellington: Geological Society of NZ Guidebook 5, 32 pp.

Davidson, J. (1987). Marks on a landscape: Auckland's volcanic cones. In J. Wilson (ed.), *The past today: Historic places in New Zealand*. Auckland: New Zealand Historic Places Trust/Pacific Publishers.

Hayward, B. W. (1987). *Granite and marble: A guide to building stones in New Zealand*. Wellington: Geological Society of NZ Guidebook 8, 56 pp.

Simmons, D. (1987). *Māori Auckland*. Auckland: The Bush Press, 96 pp.

Institute of Geological and Nuclear Sciences (1988). *Volcanic landscape: A guide to the Auckland Volcanic Field*, fold-out colour brochure, BR6.

Cox, G. J. (1989, 2000). *Fountains of fire: The story of Auckland's volcanoes*. Auckland: Collins, 28 pp.

Fairfield, G. (1992). *Pigeon Mountain o Huiarangi: The birth and death of a volcano*. Auckland: Tamaki Estuary Protection Society, 83 pp.

Kermode, L. O. (1992). *Geology of the Auckland urban area*. Institute of Geological and Nuclear Sciences, Geological Map 2. 1:50 000. Lower Hutt: Institute of Geological and Nuclear Sciences.

Smith, I. E. M., & Allen, S. R. (1993). *Volcanic hazards of the Auckland Volcanic Field* (Volcanic Hazards Information Series 5). Wellington: Ministry of Civil Defence, 34 pp.

Cobb, J. (1994). *Cornwall Park: The story of a man's vision*. Auckland: Cornwall Park Trust Board, 32 pp.

Hayward, B. W., & Gill, B. J. (eds) (1994). *Volcanoes and giants*, exhibition catalogue. Auckland: Auckland Museum, 48 pp.

Cameron, E. K., Hayward B. W., & Murdoch, G. D. (1997, 2008). *A Field Guide to Auckland: Exploring the region's natural and historic heritage*. Auckland: Godwit, 285 pp.

Chapple, G. (1999). *North Head: The enigma*. Auckland: Department of Conservation, Auckland Conservancy, 13 pp.

Cox, G. J., & Hayward, B. W. (1999). *The restless country: Volcanoes and earthquakes of New Zealand*. Auckland: HarperCollins, 64 pp.

Homer, L. L., Moore, P. R., & Kermode, L. O. (2000). *Lava and strata: A guide to the volcanoes and rock formations of Auckland*. Auckland: Landscape Publications, 96 pp.

Stone, R. C. J. (2001). *From Tamaki-Makau-Rau to Auckland*. Auckland: Auckland University Press, 342 pp.

Wilcox, M. D. (ed.) (2007). *Natural history of Rangitoto Island*. Auckland: Auckland Botanical Society, 192 pp.

Auckland Regional Council (2010). *Auckland's volcanic heritage*. Auckland, 30 pp.

Hayward, B. W., Murdoch, G., & Maitland, G. (2011). *Volcanoes of Auckland: The essential guide*. Auckland: Auckland University Press, 234 pp.

Johnston, M., & Nolden, S. (2011). *Travels of Hochstetter and Haast in New Zealand, 1858–1860*. Nelson: Nikau Press, 336 pp.

Nolden, S., & Nolden, S. B. (2013). *Hochstetter Collection Basel, Part 3 – New Zealand maps & sketches*. Auckland: Mente Corde Manu, 127 pp.

Friends of Maungawhau (2014). *Maungawhau: A short history of volunteer action*. Auckland, 107 pp.

Hayward, B. W. (2017). *Out of the ocean, into the fire: History in the rocks, fossils and landforms of Auckland, Northland and Coromandel* (Miscellaneous Publication 146). Wellington: Geoscience Society of New Zealand, 336 pp.

Useful websites

DEVORA, www.devora.org.nz/

GeoNet monitoring site and information, www.geonet.org.nz/

Auckland Emergency Management, www.aucklandemergencymanagement.org.nz/

Wikipedia, https://en.wikipedia.org/wiki/Auckland_volcanic_field

Mt Māngere field guide, Geological Society of New Zealand, www.gsnz.org.nz/html/mangere/gsfg1.htm

Scientific references used in writing this book

Affleck, D. K., Cassidy, J., & Locke, C. A. (2001). Te Pou Hawaiki volcano and pre-volcanic topography in central Auckland: Volcanological and hydrogeological implications. *New Zealand Journal of Geology and Geophysics, 44*, 313–321.

Agustín-Flores, J., Németh, K., Cronin, S. J., Lindsay, J. M., Kereszturi, G., Brand, B. D., & Smith, I. E. M. (2014). Phreatomagmatic eruptions through unconsolidated coastal plain sequences, Maungataketake, Auckland Volcanic Field (New Zealand). *Journal of Volcanology and Geothermal Research, 276*, 46–63.

Agustín-Flores, J., Németh, K., Cronin, S. J., Lindsay, J. M., & Kereszturi, G. (2015). Construction of the North Head (Maungāuika) tuff cone: A product of Surtseyan volcanism, rare in the Auckland Volcanic Field, New Zealand. *Bulletin of Volcanology, 77*, 11.

Brand, B. D., Gravley, D. M., Clarke, A. B., Lindsay, J. M., Bloomberg, S. H., Agustín-Flores, J., & Németh, K. (2014). A combined field and numerical approach to understanding dilute pyroclastic density current dynamics and hazard potential: Auckland Volcanic Field, New Zealand. *Journal of Volcanology and Geothermal Research, 276*, 215–232.

Cassata, W. S., Singer, B. S., & Cassidy, J. (2008). Laschamp and Mono Lake geomagnetic excursions recorded in New Zealand. *Earth and Planetary Science Letters, 268*, 76–88.

Cassidy, J. (2006). Geomagnetic excursion captured by multiple volcanoes in a monogenetic field. *Geophysical Research Letters, 33*, L21310.

Cassidy, J., France, S. J., & Locke, C. A. (2007). Gravity and magnetic investigation of maar volcanoes, Auckland Volcanic Field. *New Zealand Journal of Volcanology and Geothermal Research, 159*, 153–163.

Crossley P. C. (2014). Auckland Lava Caves. *New Zealand Speleological Bulletin, 11(208)*, 190–247.

Daymond-King, P., & Hayward, B. W. (2015). Just 600 years of erosion on Rangitoto's coast. *Geocene, 12*, 5–7.

Eade, J. (2009). Petrology and correlation of lava flows from the central part of the Auckland Volcanic Field. Unpublished MSc thesis, University of Auckland.

Eccles, J. D., Cassidy, J., Locke, C. A., & Spörli, K. B. (2005). Aeromagnetic imaging of the Dun Mountain Ophiolite Belt in northern New Zealand: Insight into the fine structure of a major SW Pacific terrane suture. *Journal of the Geological Society, 162*, 723–735.

Firth, C. W. (1967). Water supply of Auckland, New Zealand. Auckland Regional Authority.

Firth, J. C. (1874). Deep sinking in the lava beds of Mt Eden. *Transactions of New Zealand Institution, 7*, 460–464.

Hayward, B. W. (2008). Ash Hill Volcano, Wiri. *Geocene, 3*, 8–9.

Hayward, B. W. (2013). Volcanoes recognised by Hochstetter on the Māngere Lowlands. *Geocene, 9*, 21–23.

Hayward, B. W. (2015). Cornwallis Formation blocks in Waitomokia tuff ring deposits, Māngere. *Geocene, 12*, 2–4.

Hayward, B. W. (2015). Understanding Maungakiekie/One Tree Hill Volcano – source of its lava flows. *Geocene, 12*, 16–21.

Hayward, B. W. (2017). Eruption sequence of Rangitoto Volcano, Auckland. *Geoscience Society of New Zealand Newsletter, 23*, 4–10.

Hayward, B. W. (2018). Cemetery Crater revealed. *Geocene, 16*, 17–18.

Hayward, B. W., & Carr, G. (2014). Understanding Maungawhau/Mt Eden Volcano. *Geocene, 11*, 8–12.

Hayward, B. W., Daymond-King, P., & Daymond-King, R. (2008). North Head volcano. *Geocene, 3*, 3–4.

Hayward, B. W., & Grenfell, H. R. (2013). Did Rangitoto erupt many times? *Geoscience Society of New Zealand Newsletter, 11*, 5–8.

Hayward, J. J., & Hayward, B. W. (1995). Fossil forests preserved in volcanic ash and lava at Ihumatao and Takapuna, Auckland. *Tane, 35*, 127–142.

Hayward, B. W., Hopkins, J. L., & Smid, E. R. (2016). Māngere Lagoon predated Māngere Mt. *Geocene, 14*, 4–5.

Hayward, B. W., & Kenny, J. A. (2009). Three Kings lava lake, Auckland. *Geocene, 4*, 9–11.

Hayward, B. W., Kenny, J. A., & Grenfell, H. R. (2011). More volcanoes recognised in Auckland Volcanic Field. *Geoscience Society of New Zealand Newsletter, 5*, 11–16.

Hayward, B. W., Kenny, J. A., & Grenfell, H. R. (2012). Puhinui Craters. *Geocene, 8*, 14–18.

Hayward, B. W., Kenny, J. A., High, R., & France, S. (2011). Grafton Volcano. *Geocene, 6*, 12–17.

Hayward, B. W., & Kenny, J. A. (2013). The Royal Oak Craters. *Geocene, 9*, 15–20.

Hayward, B. W., Morley, M. S., Sabaa, A. T., Grenfell, H. R., Daymond-King, R., Molloy, C., Shane, P. A. R., & Augustinus, P. A. (2009). Fossil record of the post-glacial marine breaching of Auckland's volcanic maar craters. *Records of Auckland Institute and Museum, 45*, 73–99.

Hopkins, J. L., Wilson, C. J. N., Millet, M.-A., Leonard, G. S., Timm, C., McGee, L. E., Smith, I. E. M., & Smith, E. G. C. (2017). Multi-criteria correlation of tephra deposits to source centres applied in the Auckland Volcanic Field, New Zealand. *Bulletin of Volcanology, 79*, 55.

Horrocks, M., Augustinus, P., Deng, Y., Shane, P., & Andersson, S. (2005). Holocene vegetation, environment and tephra recorded from Lake Pupuke, Auckland, New Zealand. *New Zealand Journal of Botany, 43*, 211–221.

Houghton, B. F., Wilson, C. J. N., & Smith, I. E. M. (1999). Shallow-seated controls on styles of explosive volcanism: A case study from New Zealand. *Journal of Volcanology and Geothermal Research, 91*, 97–120.

Kenny, J. A., Lindsay, J. M., & Howe, T. M. (2012). Post-Miocene faults in Auckland: Insights from borehole and topographic analysis. *New Zealand Journal of Geology and Geophysics, 55*, 323–343.

Kereszturi, G., Németh, K., Cronin, J. S., Agustín-Flores, J., Smith, I. E. M., & Lindsay, J. (2013). A model for calculating eruptive volumes for monogenetic volcanoes – implication for the Quaternary Auckland Volcanic Field. *New Zealand Journal of Volcanology and Geothermal Research, 266*, 16–33.

Kereszturi, G., Németh, K., Cronin, S. J., Procter, J., & Agustín-Flores, J. (2014). Influences on the variability of eruption sequences and style transitions in the Auckland Volcanic Field, New Zealand. *Journal of Volcanology and Geothermal Research, 286*, 101–115.

Kermode, L. O. (1992). *Geology of the Auckland urban area*. 1:50,000. Institute of Geological and Nuclear Sciences Geological Map 2.

Kermode, L. O., Smith, I. E. M., Moore, C. L., Stewart, R. B., Ashcroft, J., Nowell, S. B., & Hayward, B. W. (1992). *Inventory of Quaternary volcanoes and volcanic features of Northland, South Auckland and Taranaki*. Geological Society of New Zealand Miscellaneous Publication, 61.

Leonard, G. S., Calvert, A. T., Hopkins, J. L., Wilson, C. J. N., Smid, E., Lindsay, J., & Champion, D. (2017). High precision 40Ar/39Ar dating of Quaternary basalts from Auckland Volcanic Field, New Zealand, with implications for eruption rates and paleomagnetic correlations. *Journal of Volcanology and Geothermal Research, 343*, 60–74.

Lindsay, J. M., Leonard, G. S., Smid, E. R., & Hayward, B. W. (2011). Age of the Auckland Volcanic Field: A review of existing data. *New Zealand Journal of Geology and Geophysics, 54*, 379–401.

Linnell, T., Shane, P. A., Smith, I. E. M., Augustinus, P., Cronin, S., Lindsay, J., & Maas, R. (2016). Long-lived shield volcanism within a monogenetic basaltic field: The conundrum of Rangitoto volcano. *New Zealand Geological Society of America Bulletin, 128*, 1160–1172.

Lowe, D. J., Shane, P. A. R., de Lange, P. J., & Clarkson, B. D. (2017). *Rangitoto Island field trip, Auckland*. Geoscience Society of New Zealand Miscellaneous Publication, 147B.

Marra, M. J., Alloway, B. V., & Newnham, R. M. (2006). Paleoenvironmental reconstruction of a well-preserved stage 7 forest sequence catastrophically buried by basaltic eruptive deposits, northern New Zealand. *Quaternary Science Reviews, 25*, 2143–2161.

McGee, L. E., Beier, C., Smith, I. E. M., & Turner, S. (2011). Dynamics of melting beneath a small-scale basaltic system: A U–Th–Ra study from Rangitoto volcano, Auckland Volcanic Field, New Zealand. *Contributions to Mineralogy and Petrology, 162*, 547–563.

McGee, L. E., Millet, M.-A., Beier, C., Smith, I. E. M., & Lindsay, J. M. (2015). Mantle heterogeneity controls on small-volume basaltic eruption characteristics. *Geology, 43*, 551–554.

McGee, L. E., Millet, M.-A., Smith, I. E. M., Németh, K., & Lindsay, J. M. (2012). The inception and progression of melting in a monogenetic eruption: Motukorea volcano, the Auckland Volcanic Field, New Zealand. *Lithos, 155*, 360–374.

Molloy, C., Shane, P., & Augustinus, P. (2009). Eruption recurrence rates in a basaltic volcanic field based on tephra layers in maar sediments: Implications for hazards in the Auckland Volcanic Field. *Geological Society of America Bulletin, 121*, 1666–1677.

Needham, A. J., Lindsay, J. M., Smith, I. E. M., Augustinus, P., & Shane, P. A. (2011). Sequential eruption of alkaline and subalkaline magmas from a small monogenetic volcano in the Auckland Volcanic Field, New Zealand. *Journal of Volcanology and Geothermal Research, 201*, 126–142.

Németh, K., Agustín-Flores, J., Briggs, R., Cronin, S. J., Kereszturi, G., Lindsay, J. M., Pittari, A., & Smith, I. E. M. (2012). *Field guide: Monogenetic volcanism of the South Auckland and Auckland Volcanic Fields. 4th International Maar Conference, Auckland, New Zealand.* Geoscience Society of New Zealand Miscellaneous Publication, 131B.

Németh, K., Cronin, S. J., Smith, I. E. M., & Agustín-Flores, J. (2012). Amplified hazard of small volume monogenetic eruptions due to environmental controls, Orakei Basin, Auckland Volcanic Field, New Zealand. *Bulletin of Volcanology, 74*, 2121–2137.

Nowak, J. F. (1995). Lava flow structures of a basaltic volcano, Rangitoto Island, Auckland, New Zealand. Unpublished MSc thesis, University of Auckland.

Searle, E. J. (1959). The volcanoes of Ihumatao and Māngere, Auckland. *New Zealand Journal of Geology and Geophysics, 2*, 870–888.

Searle, E. J. (1961). Volcanoes of the Ōtāhuhu–Manurewa district, Auckland. *New Zealand Journal of Geology and Geophysics, 4*, 239–255.

Searle, E. J. (1962). The volcanoes of Auckland City. *New Zealand Journal of Geology and Geophysics, 5*, 193–227.

Searle, E. J. (1965). Auckland Volcanic District. In B. N. Thompson & L. O. Kermode, New Zealand Volcanology, Northland, Coromandel, Auckland. *DSIR Information Series, 49*, 90–103.

Shane, P. A., & Hoverd, J. (2002). Distal record of multi-sourced tephra in Onepoto Basin, Auckland, New Zealand: Implications for volcanic chronology, frequency and hazards. *Bulletin of Volcanology, 64*, 441–454.

Shane, P. A. R., & Zawalna-Geer, A. (2011). Correlation of basaltic tephra from Mt Wellington volcano: Implications for the penultimate eruption from the Auckland Volcanic Field. *Quaternary International, 246*, 374–381.

Spörli, K. B., Black, P. M., & Lindsay, J. M. (2015). Excavation of buried Dun Mountain–Maitai terrane ophiolite by volcanoes of the Auckland Volcanic Field, New Zealand. *New Zealand Journal of Geology and Geophysics, 58*, 229–253.

Zawalna-Geer, A., Lindsay, J. M., Davies, S., Augustinus, P., & Davies, S. (2016). Extracting a primary Holocene cryptotephra record from Pupuke maar sediments, Auckland, New Zealand. *Journal of Quaternary Science, 31*, 442–457.

Acknowledgements

I am forever grateful to the late Les Kermode for sharing his extensive knowledge of Auckland's volcanoes with me during the 1990s as we co-led Auckland Geology Club field trips to all parts of the Auckland Volcanic Field.

This book is founded on our 2011 book *Volcanoes of Auckland: The Essential Guide*. I am grateful to co-author Graeme Murdoch who has allowed me to repeat much of his input on Māori history and names that we published in that book. Once again, I thank all those who provided specialist advice and checked the accuracy of the earlier book.

In the past eight years there has been a continuing high level of research on Auckland's volcanoes, mainly under the auspices of the DEVORA (Determining Volcanic Risk in Auckland) project – a collaboration between staff and students of the University of Auckland, Massey University, Victoria University of Wellington, Canterbury University and GNS Science. Their research has greatly enhanced our understanding of the potential volcanic hazard faced by Auckland but, most of all, the work of Graham Leonard and Jenni Hopkins has revolutionised our knowledge of the ages of most of the field's volcanoes.

In addition, for specialist help and assistance with this revised edition, I thank Peter Crossley, Ilmars Gravis, Jenni Hopkins, Jill Kenny, Graham Leonard, Jan Lindsay, Brian McDonnell, Karoly Németh, Sascha Nolden, Elspeth Orwin, Elaine Smid and Dave Towns. Hugh Grenfell, Jo Horrocks, Nick Horspool and the late Margaret Morley are also thanked for allowing reproduction of several of their superb figures.

Photographs, maps and figures are by the author except where credited to others or a historic source. I thank Hugh Grenfell, Jill Kenny, the late Les Kermode, Lucy McGee, Phil Shane and Elaine Smid who have kindly allowed me to reproduce their photographs. I acknowledge the New Zealand GeoNet project and its sponsors the Earthquake Commission, GNS Science and Land Information New Zealand, and the Ministry of Civil Defence and Emergency Management for providing images. Jill Kenny expertly used her computer skills to magically remove blemishes and enhance some of the more difficult older photographs. I especially thank David Fraser for his unfailing enthusiasm to discover and share historic photographs and information about the southern volcanoes in the field. As always, Alastair Jamieson has been a rock star in organising the helicopter flight for our updated aerial photography of some volcanoes and in supplying so many splendid shots.

The AUP team Sam Elworthy, Katharina Bauer, Sophia Broom, Elizabeth Newton-Jackson, copy-editor Louise Belcher, proofreader Matt Turner and designer Carolyn Lewis are thanked for their friendly encouragement and support throughout all stages of the book's genesis. David Veart is thanked for his insightful review.

Finally, I thank my wife Glenys, and daughters Kathryn, Jessica and Clare, along with members of the Auckland Geology Club, who have been dragged uncomplainingly around Auckland's volcanoes for many years and often asked to stand in as scales for photographs.

Index

Bolded entries are to occurrence in maps, figures or photos